MW00838089

BLUEPRINT READING FOR HEATING, VENTILATING, AND AIR CONDITIONING

Delmar Publishers' Online Services

To access Delmar on the World Wide Web, point your browser to:
http://www.delmar.com/delmar.html
To access through Gopher: gopher://gopher.delmar.com
(Delmar Online is part of "thomson.com", an Internet site with information on
more than 30 publishers of the International Thomson Publishing organization.)
For information on our products and services:
email: info@delmar.com
or call 800-347-7707

BLUEPRINT READING FOR HEATING, VENTILATING, AND AIR CONDITIONING

Frank C. Miller

Wilma B. Miller

Delmar Publishers

I(T)P An International Thomson Publishing Company

Albany • Bonn • Boston • Cincinnati • Detroit • London • Madrid
Melbourne • Mexico City • New York • Pacific Grove • Paris • San Francisco
Singapore • Tokyo • Toronto • Washington

NOTICE TO THE READER

Cover Credit: Michael Prinzo

Delmar Staff

Acquisitions Editor: Paul Shepardson
Editorial Assistant: Dawn Daugherty
Project Editor: Patricia Konczeski

Production Coordinator: Toni Bolognino
Art/Design Coordinator: Michael Prinzo

COPYRIGHT © 1996
Delmar is a division of Thomson Learning. The Thomson Learning logo is a registered trademark used herein under license.

Printed in the United States of America
3 4 5 6 7 8 9 10 XXX 01 00

For more information, contact Delmar, 3 Columbia Circle, PO Box 15015, Albany, NY 12212-0515;
or find us on the World Wide Web at http://www.delmar.com

· **International Division List**

Japan:
Thomson Learning
Palaceside Building 5F
1-1-1 Hitotsubashi, Chiyoda-ku
Tokyo 100 0003 Japan
Tel: 813 5218 6544
Fax: 813 5218 6551

Australia/New Zealand
Nelson/Thomson Learning
102 Dodds Street
South Melbourne, Victoria 3205
Australia
Tel: 61 39 685 4111
Fax: 61 39 685 4199

UK/Europe/Middle East:
Thomson Learning
Berkshire House
168-173 High Holborn
London
WC1V 7AA United Kingdom
Tel: 44 171 497 1422
Fax: 44 171 497 1426

Latin America:
Thomson Learning
Seneca, 53
Colonia Polanco
11560 Mexico D.F. Mexico
Tel: 525-281-2906
Fax: 525-281-2656

Canada:
Nelson/Thomson Learning
1120 Birchmount Road
Scarborough, Ontario
Canada M1K 5G4
Tel: 416-752-9100
Fax: 416-752-8102

Asia:
Thomson Learning
60 Albert Street, #15-01
Albert Complex
Singapore 189969
Tel: 65 336 6411
Fax: 65 336 7411

Library of Congress Cataloging-in-Publication Data
Miller, Frank C.
    Blueprint reading for heating, ventilating, and air conditioning :
instructors guide / Frank C. Miller, Wilma B. Miller.
      p.   cm.
    Includes index.
    ISBN 0-8273-6872-0
    1. Heating—Equipment and supplies—Drawings. 2. Ventilation—
-Equipment and supplies—Drawings. 3. Air conditioning—Equipment
and supplies—Drawings. 4. Blueprints. I. Miller, Wilma B.
II. Title.
TH7345.M555   1995   95-1920
697—dc20   CIP

CONTENTS

PREFACE

This text is written for the student in a heating, ventilating, and air conditioning (HVAC) program of a community college or technical school, who plans a career in the building trades. The HVAC technician must be a "jack-of-all-trades" on the job site and in the office. This textbook, therefore, addresses various areas of expertise that may be expected, including the following:

(1) reading and interpreting architectural, plumbing, HVAC, and electrical plans for the sizing and installation of equipment;

(2) being able to use various kinds of scales and measuring devices, and recognizing standard symbols and abbreviations;

(3) following safety procedures in the office and on the job site;

(4) understanding, interpreting, and being able to create isometric, orthographic, and oblique drawings;

(5) working with blueprints for ductwork, piping, electrical wiring, controls, and equipment layout;

(6) being able to do freehand sketching and drafting as needed in the office, shop, or at the job site;

(7) recognizing the relationships and responsibilities of owners, architects, engineers, designers, drafters, contractors, and others involved in the planning and construction of any building;

(8) knowing how to operate the various types of printmaking equipment in the office, and understanding why some types of prints are better than others for specific uses;

(9) recognizing the advantages of using Computer-Aided Design (CAD) or manual drafting equipment to create plans; and

(10) realizing that the HVAC system is an integral part of the total structure, requiring coordination with the other trades and design personnel.

Numerous illustrations, photographs, and actual building plans have been included for the student to refer to as needed. Attention has been given to understandable vocabulary, precise sentences, stated objectives and summaries for each chapter, and review questions and exercises that are realistic and related to current practices in the field. The appendices furnish tables of easy-to-use reference material. The *Instructor's Guide* contains suggestions for numerous enrichment activities in addition to those offered in the textbook.

The organization of this text is logical and practical. Part I begins with the elements of blueprint reading, progressing from a review of mathematical functions, measurement and scales, and symbols and abbreviations, to the process of creating and interpreting working drawings and construction drawings. The components of these drawings, including details, sections, elevations, specifications, and title blocks, are explained and illustrated in separate chapters. Discussion is included on the roles played by the owner, architect, engineer, designer, drafter, contractor, and others involved in the building process.

Part II deals with freehand sketching and drafting. Lines, lettering, shading, and crosshatching are covered in separate chapters. Orthographic projection, isometric drawings, and oblique drawings are discussed and illustrated in detail. Practical suggestions are given about organizing drafting plans, drawing construction elevations, sketching, and manual drafting.

Part III contains numerous exercises for the student, using an adaptation of actual construction plans. The student can read and interpret architectural, mechanical, electrical, plumbing, and ductwork plans just as a technician might be expected to do in the office or on the job site.

The text is organized to allow the student (or instructor) to work straight through or to rearrange the order of subject matter as needed. Chapter 1 is a self-test to assess basic mathematical skills needed for the subject matter in the text. The chapters on measurement and on reading scales may seem over-simplified, but the authors and other instructors have noticed that average students often have difficulty in working with scales and measurement.

ACKNOWLEDGMENTS

The authors wish to acknowledge the contributions of many people who have helped in the preparation of this book, including the following: R.L. Rash for technical assistance, for support, and for supplying and preparing architectural plans; Greg C. Miller for help with the sections on symbols and lettering, as well as assistance with computer-assisted drafting; Bill Johnson for photography and consultation during the writing of the text; Margaret F. Miller for suggestions for the Self-Test and technical advice; Laura Miller for scanning and retouching photographs and for help with art work; W. R. Harris for furnishing information on specifications; Jeff and Jenny Miller, Eleanor Harris and other family members and friends who have offered encouragement and proofreading assistance during the writing of this book.

We appreciate the help of the following companies that supplied materials and photographs:

Hewlett-Packard Company, San Diego, California

Blu-Ray, Incorporated, Essex, Connecticut

Central Piedmont Community College Bookstore

Duncan-Parnell, Inc., Charlotte, North Carolina

We wish to thank Kim Davies, Patrick Delaney, Mark Huth, and other staff members of Delmar Publishers for their assistance. We are also grateful to the following educators who reviewed the manuscript and offered suggestions for improvement:

Paul Ackerman, Texas State Technical College

John Barry, International Union of Operating Engineers

Pat Courington, Mid-Florida Technical Institute

Harold L. Eldred, Northeast Metro Tech College

Jim Johnson, ABC Tech

Joseph P. Moravek, Lee College

John N. Morris, Henry Ford Community College

PART I

BLUEPRINT READING

REVIEW OF BASIC MATHEMATICAL PROCEDURES

OBJECTIVES

OBJECTIVES

After studying this chapter, you should be able to:

- discuss reasons why heating and air-conditioning technicians must know basic mathematical procedures
- determine what areas of mathematics you need to review before proceeding with this text
- use equations to solve for unknown information

1.1 INTRODUCTION

The successful heating, ventilating, and air-conditioning (HVAC) technician must have a working knowledge of basic mathematical procedures. The technician is called on daily to accurately add, subtract, multiply, and divide whole numbers, fractions, and decimals. Taking measurements on the job requires the ability to calculate the area and volume of rectangles, triangles, and circles. The volume of cylinders and other geometric figures must often be determined. To come up with accurate answers, the technician must be able to formulate and use basic equations.

Self-Test

The following problems are to be used as a self-test of the student's ability to use basic mathematical procedures. The student is urged to work through all of the problems before checking the answers. The correct answers to the odd-numbered problems are listed at the end of this chapter. The correct answers to the even-numbered problems are shown in the *Instructor's Guide*.

Formulas for solving problems in Chapter 1 appear in Appendix A at the end of this textbook. The student should use Appendix A as a reference as needed.

Students who have difficulty answering any part of the self-test correctly are encouraged to review and learn those procedures where a weakness is evident before continuing with succeeding chapters in this text. This review can be accomplished in any of several ways. The individual may need only to refer to a basic mathematics textbook to refresh the memory. A computer program for self-study may be available to re-teach (or teach for the first time) those areas that cause difficulties. A general mathematics course may be well worth the time.

The point is that heating and air-conditioning technicians today need more than good common sense and the desire to do hands-on work in the mechanical field. Technicians are often expected to calculate prices, estimate amounts of materials needed, figure sizes of ductwork and other equipment, and carry on a host of other procedures that require a sound foundation in basic mathematical procedures and calculations.

The self-test begins with the basic functions of addition, subtraction, multiplication, and division, and it progresses into problems of increasing complexity. Remember, careless mistakes can be as costly as those resulting from ignorance of formulas and other procedures.

1.2 ADDING, SUBTRACTING, MULTIPLYING, AND DIVIDING WHOLE NUMBERS

1. 236
 + 98

2. 5490
 + 436

3. 567
 − 421

4. 6783
 − 2357

5. 2896
 x 326

6. 242
 x 32

7. 18)216

8. 25)400

9. The air quantities delivered to three areas of a building are as follows: 988 cubic feet per minute (CFM), 2349 CFM, and 6000 CFM. What is the total air quantity (in CFM) delivered to the three areas?

10. An oil company delivers the following quantities of oil to a business which has three tanks: 5689 gallons in Tank 1, 528 gallons in Tank 2, and 100 gallons in Tank 3. What is the total number of gallons of oil delivered to all three tanks?

11. A contractor has two houses under construction. The first house has a square footage of 4795 square feet and the second house has a square footage of 2357 square feet. How much larger is the first house than the second one?

12. A sheet metal shop was billed for 2387 pounds of sheet metal. The shop foreman says that only 2298 pounds of sheet metal was delivered. If the foreman is correct, how many pounds of sheet metal would be needed to complete the order correctly?

13. If 1296 feet of copper tubing will be needed for the plumbing for one building, how many feet of copper tubing will be needed for 15 identical buildings?

14. A rectangular plot of land measures 567 feet frontage and 1234 feet deep. What is the square footage of the plot of land?

15. Two hundred air filters, sized 18 inches by 24 inches, are needed for a job. The filters are packed in boxes of 10 filters per box. How many boxes of filters should be ordered for the job?

16. The time estimated for completing a drafting job is 385 hours. If a drafter works 35 hours per week on this job, how many weeks will be needed to complete the job?

1.3 FRACTIONS

A. Convert these improper fractions into mixed numbers:

17. $^{27}/_5$ **18.** $^4/_3$ **19.** $^{18}/_4$ **20.** $^5/_4$

B. Use the correct procedures to answer the following:

21. $^1/_5 + ^4/_{15} + ^7/_{15} =$

22. $^2/_3 + ^1/_4 + ^3/_6 =$

23. $^{19}/_{24} − ^3/_8 =$

24. $2^1/_4 − ^7/_8 =$

25. $^1/_2 \times ^1/_2 =$

26. $^5/_6 \times ^{11}/_{12} =$

27. $^1/_3 \div 2 =$

28. A HVAC technician measures the components of a wall as follows: 4 inches of brick, $^3/_8$ inches of sheathing, $3^1/_2$ inches of insulation, and $^1/_2$ inch of sheetrock. What is the total thickness of the wall?

29. A gasket on a walk-in cooler must be replaced. The door of the cooler measures $34^1/_2$ inches wide and $71^1/_4$ inches high. How many inches of gasket material will be needed to fit around the outside edges of the cooler door?

30. If $^1/_8$ inch is cut from a section of tubing $1^1/_2$ inches long, how much tubing is left?

31. A cylinder containing 19 pounds of R-22 refrigerant is used to recharge three air-conditioning units. Two pounds are used to fill the first unit, $3^1/_2$ pounds are used for the second unit, and $2^1/_4$ pounds are used for the third unit. How many pounds are left in the cylinder?

32. An oil burner uses $^1/_2$ gallon of fuel per hour. How many gallons of oil will be used over a period of $7^1/_2$ hours?

33. A HVAC technician spends $3^1/_2$ hours repairing two units. If he charges $15.00 per hour for labor, how much will he charge for time spent repairing the two units?

34. A service technician spent $7^1/2$ hours repairing 3 units. If he spent an equal amount of time on each unit, how many hours did it take to repair each unit?

35. A steel pipe $9^3/4$ feet long is to be divided into 3 equal sections. How long will each section be?

36. Into how many $^1/4$ inch sections can a line $^9/16$ inches long be divided?

1.4 DETERMINING THE PERIMETER OF A SQUARE OR RECTANGLE

A. Find the perimeter of each of the following squares:

37. Each side is 6 feet wide

38. Each side is 4 inches wide

B. Find the perimeter of rectangles with the following dimensions:

39. Side A = 10 inches
Side B = 12 inches
Perimeter =

40. Side A = 6 feet
Side B = 7 feet
Perimeter =

1.5 DETERMINING THE AREA OF A SQUARE OR RECTANGLE

A. Find the area of each of the following squares:

41. Each side is nine meters wide

42. Each side is thirty inches wide.

B. Find the area of rectangles with the following dimensions:

43. Side A = 42 feet
Side B = 81 feet
Area =

44. Side A = 30 inches
Side B = 24 inches
Area =

1.6 DETERMINING THE AREA OF A TRIANGLE

A. Find the area of triangles with the following dimensions:

45. Base = 5 feet
Height = 8 feet
Area =

46. Base = $4^1/2$ inches
Height = $9^1/2$ inches
Area =

47. Base = 21 feet
Height = 50 feet
Area =

48. Base = 17 inches
Height = 17 inches
Area =

1.7 DETERMINING THE AREA OF A CIRCLE

Use the proper equation to determine the area of a circle if the known dimensions are as follows:

49. Radius = 4 feet

50. Radius = $2^1/2$ inches

51. Diameter = 7 inches

52. Diameter = 40 feet

1.8 DETERMINING THE CIRCUMFERENCE OF A CIRCLE

Use the proper equation to determine the circumference of each of the following circles:

53. Radius = 3 inches

54. Diameter = 12 inches

1.9 DETERMINING VOLUME

Find the volume of each figure described below.

55. Cylindrical tank with diameter of 3 feet and length of 5 feet

56. Cylindrical tank with diameter of 1 foot and length of 10 feet

57. Rectangular tank 3 feet wide, 4 feet high, and 8 feet long

58. Rectangular tank 10 feet wide, 7 feet high, and 8 feet long

59. Elliptical tank having a small radius of 2 feet, a large radius of 4 feet, and a length of 5 feet

60. Triangular tank with a 5-foot base, 5 feet long, and 5 feet high (the sloped sides are equal)

1.10 EQUATIONS

Write the equation for each of the following:

61. Area of a square

62. Area of a circle

63. Perimeter of a rectangle

64. Area of a rectangle

65. Area of a triangle

66. Circumference of a circle

67. Volume of a cylinder

68. Volume of a rectangular container

1.11 ADDITIONAL PROBLEMS

69. A full oil tank holds 300 gallons of Grade 2 fuel oil. During the past three months the amounts of oil burned are 21 gallons, 55 gallons, and 127 gallons. How much oil is left in the tank?

70. One roll of duct insulation insulates 30 linear feet of duct. How many linear feet can be insulated with 65 rolls?

71. There are 144 electrical connectors in a box. How many connectors are there in 9 boxes?

72. A fuel tank holds 200 pounds of fuel. One gallon of fuel weighs 7.1 pounds. How many gallons of fuel are in a full tank?

73. A cylindrical oil tank is 3 feet in diameter and 5 feet long. How many cubic feet of oil will this tank hold?

74. A 100-foot coil of $^3/_8$-inch copper tubing weighs 19.8 pounds. How much does one foot of this tubing weigh?

75. A stack of 80 pieces of sheet metal weighs 650 pounds. What is the weight per sheet?

ANSWERS TO SELF-TEST QUESTIONS

(Answers to odd-numbered questions follow. Even-numbered questions are answered in Instructor's Guide.)

1. 334

3. 146

5. 944,096

7. 12

9. 9337 CFM

11. 2438 square feet

13. 19,440 feet

15. 20

17. $5^2/_5$

19. $4^1/_2$

21. $^{14}/_{15}$

23. $^{10}/_{24}$ or $^5/_{12}$

25. $^1/_4$

27. $^1/_6$

29. $211^1/_2$ inches

31. $11^1/_4$ pounds

33. $52.50

35. $3^1/_4$ feet

37. 24 feet

39. 44 inches

41. 81 square meters

43. 3402 square feet

45. 20 square feet

47. 525 square feet

49. 50.24 square feet

51. 38.465 square inches

53. 18.84 inches

55. 35.325 cubic feet

57. 96 cubic feet

59. 125.6 cubic feet

61. A = L x W or A = 4 x S

63. P = 2L + 2W

65. A = L x H ÷ by 2

67. V = 3.14 x r x r x H

69. 97 gallons

71. 1296 connectors

73. 35.325 cubic feet

75. 8.125 pounds

2

SAFETY

OBJECTIVES

After studying this chapter, you should be able to:

- discuss safety precautions for technicians working in the blueprint room and the drafting room
- discuss safety precautions for technicians on the job site
- tell why safety precautions are necessary for persons working with refrigerants and hazardous materials
- explain why safeguards are needed for persons working with electrical power
- explain what the acronym OSHA means, and discuss the impact of OSHA on safety in the workplace

2.1 INTRODUCTION

In spite of an emphasis on safety in the workplace, U.S. workers suffer approximately two million disabling accidents on the job annually, with some 80 million workdays lost each year. Most employers realize that safety is good business, and that profits are greater when employees are safe and healthy. Yet when owners see the same conditions and practices every day, safety problems become commonplace, and there may be no great urgency to make corrections. Workers must share the responsibility for safe conditions by inspecting the work-

place frequently, reporting potential hazards, following manufacturers' instructions when using tools or machinery, and acting responsibly on the job.

Safety should be on a technician's mind at all times. Common sense and safety precautions must be observed in the blueprint room, in the drafting room, and on the job. The listing of safety rules in this chapter is not in any order of importance and is certainly not all-inclusive. The intent of this chapter is to remind the technician that accidents happen even in an environment not considered to be hazardous—it only takes the technician becoming careless or thoughtless on the job.

2.2 SAFETY RULES FOR THE BLUEPRINT ROOM AND DRAFTING ROOM

1. When using the blueprint machine, do not wear loose clothing, neckties, or rings that might get caught by moving parts of the machine.

2. Store blueprints and other paper products carefully in order to prevent a fire hazard.

3. Avoid breathing ammonia fumes from the printing process.

4. Provide adequate lighting in the space where plans are drawn and where blueprints are read.

5. Do not place coffee and soft drinks in the area where blueprints and drawings are being used or stored. Liquid spills can ruin plans and construction documents.

6. Avoid eating or handling food near any work space where blueprints or contract documents are exposed.

7. Refrain from "horseplay" in the drafting room.

8. Avoid getting paper cuts when working with blueprints and other documents.

2.3 SAFETY RULES FOR TECHNICIANS ON THE JOB SITE

1. Keep the work area clean and orderly.

2. Wear proper protective gear, such as gloves and safety glasses, when working in extreme temperatures,

3. Be aware that pressures above and below atmospheric pressure can cause injury. (High pressures can pierce and cut the skin or inflict damage from objects that become airborne. A vacuum can cause a blood blister on the skin.)

4. Wear gloves and safety glasses when working with refrigerants.

5. Be sure that proper ventilation is present when working with refrigerants. (Refrigerants, being heavier than air, displace air. If adequate ventilation is not present, the technician may lose consciousness or become ill.)

6. Do not store refrigerants near an open flame such as the pilot on a water heater or furnace. Refrigerant does not burn; however, when it is exposed to a flame, a toxic and corrosive gas is given off that deteriorates metal in the vicinity. This gas burns the eyes and hampers breathing.

7. When working with hand tools, use the proper tool for the job. Using the improper tool can cause injury to the technician, to the tool, and to the object being worked on.

8. Never release refrigerants into the atmosphere. Refrigerant reclaim systems are available, and federal law requires that all refrigerants be recovered rather than be released into the atmosphere to destroy the ozone layer.

9. Never pour refrigerant oil on the ground, it pollutes the earth. Send used refrigerant oils to a recycling company for treatment.

10. Transferring refrigerants from one cylinder to another should be done by an authorized person at the refrigerant manufacturing plant. Throwaway cylinders are meant to be used one time and then disposed of. It is illegal to refill these throwaway cylinders.

11. Shut off electrical power at the distribution entrance panel before installing equipment. A locking device should be put on the disconnect device by the installing technician so that someone else cannot turn the power back on inadvertently.

12. Be sure that all electrical connections to air-conditioning equipment are grounded.

13. All electric hand tools used by the technician should be grounded.

14. Take special precautions when using a torch around combustible materials. Use a protective shield to prevent the heat from damaging surrounding materials and combustibles.

15. Never try to solder or weld a pipe or tubing that is sealed. Heat causes the contents of the pipe or tubing to expand and may cause bodily injury.

16. When working around rotating equipment, do not wear neckties, gloves, and loose-fitting clothing that could become caught in the equipment, resulting in injury or death.

17. Make sure all nuts are tight on motor couplings before starting the motor.

18. Never attempt to stop a motor or other belt-driven equipment by gripping the belt.

19. Use your legs, not your back, to lift heavy objects. Always keep your back straight when lifting objects.

20. Personnel subject to lifting heavy weights should wear steel-toed shoes.

21. When using recommended chemicals in the air-conditioning field, know the chemicals, and follow the manufacturer's recommendations for use. **Do not use chemicals except as recommended by manufacturer.**

22. When working with electricity, consider every wire energized until proven de-energized. Verify that all circuits serving the working area are turned off and locked out. This is true with *any* voltage.

23. Wear protective eye glasses or shields when working with liquid sprays or objects that might become airborne.

24. When replacing a three-phase motor, verify the rotation of the motor before connecting the drive.

25. Avoid "horseplay" in the work area.

26. Be sure that flame-source heating systems have ample combustion air and are properly vented.

27. Never use an open flame where the air has heavy concentrations of refrigerants. Doing so would cause phosgene gas, a colorless respiratory irritant with an unpleasant odor, to be created.

2.4 SAFETY—FEDERAL LAW

The federal government has recognized the fact that unsafe practices in the workplace are hazardous to workers and waste valuable human and material resources. In an effort to create a safer working environment, the federal government set up the Occupational Safety and Health Administration (OSHA) in 1970 to establish a set of regulations that companies must follow.

Regulations enforced by OSHA include those dealing with protective garments, fire prevention, and exposure to asbestos, lead, and other potentially dangerous substances. OSHA tries to educate workers and employers about hazards in the workplace. OSHA representatives visit companies and manufacturing plants unannounced to check on safety conditions for the workers. OSHA can issue citations for safety violations and can levy fines on the company. These fines can be heavy enough to put a small company with significant safety violations out of business.

Although there are frequent complaints from executives about paperwork, expensive changes required that may or may not improve safety for workers, and too much governmental intrusion into private industry, OSHA continues to make a significant impact on industry. In an effort to decentralize authority and to have an even more far-reaching effect, the federal government has encouraged states to develop their own health and safety programs to be approved by OSHA. The intent of government is to make workers and management realize that safe conditions and practices must have high priority in the workplace.

SUMMARY

- The rules of safety and common sense should be followed by the technician in the office and on the job.
- Food and drink should not be present in blueprinting or drafting areas.
- Fumes and moving parts related to the printing process can be dangerous.
- Safety equipment is necessary when working in extreme temperatures or with hazardous equipment on the job.
- Proper ventilation, safe storage of materials, and precautions when working with electricity prevent hazardous conditions.
- There are numerous legal and practical considerations for persons working with refrigerants.
- A federal agency commonly known as OSHA enforces regulations dealing with safety issues.
- Safe conditions and practices should be a cooperative goal of labor, management, and government agencies.

REVIEW QUESTIONS

1. List three rules for safety in the drafting or blueprint room.

2. Explain why proper ventilation is necessary when the technician is working with refrigerants.

3. Why is it especially dangerous to store refrigerants near an open flame?

4. How should refrigerants be disposed of? Why?

5. What is the purpose of temporarily installing a locking device at the electrical distribution entrance panel?

6. List two instances in which proper clothing or protective gear is essential for safety.

7. Define OSHA, and discuss briefly three of the duties of OSHA.

8. Why are business executives sometimes critical of OSHA?

9. In your opinion, who is responsible for seeing that proper safety practices are followed in the workplace? How can a company set up an effective safety program?

3

LINEAR MEASUREMENT

OBJECTIVES

After studying this chapter, you should be able to:

■ list three reasons why technicians need to understand scales and measurements

■ define linear measurement

■ differentiate between units of measurement in the English system and the metric system

■ understand and read a twelve-inch scale

■ describe the relationship between inches, feet, and yards

■ list and define the major instruments used in linear measurement

3.1 INTRODUCTION

To be successful on the job, the HVAC technician must know how to read standard scales and take accurate measurements. Dimensions and measurements must be correct for the calculation of heating and cooling loads. Correct dimensions are also necessary for manufacturing and installing ductwork, piping, and equipment.

Scales (also called scale rules) are measuring devices or instruments that enable persons to take measurements and then repeat these exact measurements at a later time and in another location. Scales have been designed so that they can be carried on the job easily.

Examples of the most common instruments for measuring linear distances are the ruler, the folding rule, and the steel tape rule. These devices are divided into standard graduations that can be read at any time. In other words, dimensions taken on the job are recorded and then reproduced in the shop and in the drafting room.

To establish and maintain accurate measurement standards for industry, science, and commerce in the United States, the National Bureau of Standards was established in 1901. This agency compares and coordinates its standards with those of other countries in the world. All scales and measuring devices have been calibrated in accordance with primary standards of the National Bureau of Standards, verifying and insuring standard units of measure in the United States and throughout the world.

3.2 THE ENGLISH SYSTEM AND THE METRIC SYSTEM

The word linear is defined as "of or pertaining to a line or lines." Linear measurement is the measurement between two points along a straight line. This measurement can be the distance from one end of an object to the other or any distance along the way. It can be the distance from one wall to the other, the width of a door, the height of a ceiling, or the width of a piece of ductwork. Any time a person uses a folding rule, a twelve-inch ruler, a steel tape, or a yardstick to measure a distance, the person is taking a linear measurement.

Two linear measuring systems, the English system and the metric system, are commonly used in the world today. The English system has as basic units of measurement the inch (in.), foot (ft.), yard (yd.), and mile (mi.). Symbols for feet and inches are often used in the building industry. Using these symbols, six feet would be written 6'-0".

The basic unit of the metric system is the meter. Prefixes express multiples of the meter. Common metric prefixes are *deci* (10), *centi* (100), and *milli* (1000). One meter contains 10 decimeters, 100 centimeters, and 1000 millimeters. Commonly used abbreviations in the metric system include the following: millimeter (mm), centimeter (cm), decimeter (dm), and meter (m).

Although most countries use the metric system, workers at all levels of the building trades in the United States have almost always preferred the English system. In 1975 Congress passed the Metric Conversion Act, calling for voluntary change from the English system to the metric system; however, very little change in that direction occurred. Only a few companies that manufacture heating and air-conditioning equipment in the United States use the metric system, and architects and engineers continue to use the English system for drawing and dimensioning plans.

Since metric dimensions are occasionally used in heating and air conditioning, metric conversion tables are included in Appendix B to help the student calculate conversions if necessary. We will use the English system throughout the body of this text.

3.3 REDUCED SCALES

Because it is not practical to draw plans full scale, the technician must be able to read and use reduced scales on construction plans. The reduced scale is used to measure distances; therefore, this kind of measurement is linear measuring. The architect's scale (Chapter 5) and the engineer's scale (Chapter 6) are used to reduce the scale of drawings.

3.4 THE ONE-INCH GRADUATION

The basic graduation for a measuring instrument in the English system is the inch. In Figure 3.1A, a

rectangle has been drawn to represent one inch. In Figure 3.1B, the inch is divided into two equal parts. Each part of the inch is one-half ($^1/_2$) inch.

In Figure 3.1C, the two $^1/_2$-inch segments have been divided into equal parts. With four equal parts of the original inch, each segment represents one-fourth (1/4) inch.

In Figure 3.1D, each $^1/_4$-inch segment has been divided in half, creating eight equal parts called one-eighth ($^1/_8$) inch segments.

On some scales the $^1/_8$-inch segments have been divided in half to form $^1/_{16}$-inch segments. The inch can be divided into 32 or 64 segments (also called a scale of $^1/_{32}$ or $^1/_{64}$). These divisions, however, are smaller than needed for most heating and air-conditioning work.

On the one-inch scale, the longest division lines represent the one-inch graduation. Slightly shorter lines represent the one-half inch graduation, even

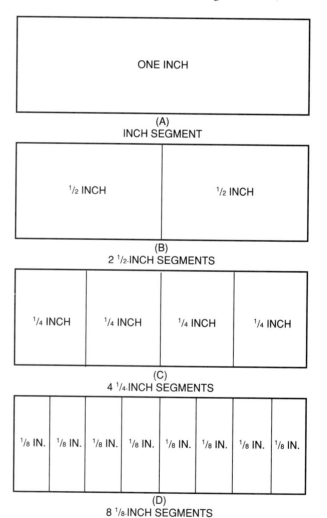

Figure 3.1 How an inch is divided

shorter lines represent the one-quarter inch graduation, and so on. By noticing the differentiation between the various lengths of division lines, one can easily determine the inch, $^1/_2$-inch, $^1/_4$-inch, $^1/_8$-inch, and $^1/_{16}$-inch graduations.

3.5 THE ONE-FOOT GRADUATION

In the English system, twelve inches are equal to one foot. When a measuring instrument is more than one foot (twelve inches) long, the markings for each foot are usually longer and more noticeable than the graduation marks for each inch (Figure 3.2).

3.6 THE ONE-YARD GRADUATION

The measuring instrument containing three one-foot graduations is called the yardstick. The one-foot graduations are subdivided into one-inch, $^1/_2$-inch, $^1/_4$-inch, and $^1/_8$-inch graduations. In most cases yardsticks are made of wood and are used in the home as a general use instrument. Yardsticks are hard to carry on the job, and they are not commonly used in the heating and air-conditioning industry.

3.7 LINEAR MEASUREMENT INSTRUMENTS

The most common linear measurement instruments used in the heating and air-conditioning industry are as follows:

One-foot scale—usually steel construction (Figure 3.3)

Six-foot folding rule—usually wood construction with metal pivoting joints every six inches (Figure 3.4)

Steel tape—flexible steel tape varying in length from a few feet to a hundred feet (Figure 3.5)

Figure 3.2 A three-foot rule

3.8 DECIMALS AND FRACTIONS

Occasionally a technician needs to be able to convert a number from a decimal to a fraction (or from a fraction to a decimal). For this reason, tables of commonly used fractions and their decimal equivalents have been included in Appendix B.

To review briefly, remember the following: To change a fraction to a decimal form, divide the numerator (top number) by the denominator (bottom number). The fraction $^2/_5$, therefore,

Figure 3.3 One-foot steel rule *(Photo by Bill Johnson)*

Figure 3.4 Six-foot folding rule *(Photo by Bill Johnson)*

Figure 3.5 Steel tapes in different lengths *(Photo by Bill Johnson)*

becomes .40 (forty hundredths) as a result of this mathematical procedure.

Changing a decimal to a fraction can also be accomplished in a few simple steps. If the decimal is stated in hundredths (.40), place the numeral 40 without the decimal over the numeral 100 ($^{40}/_{100}$). If the decimal is stated in thousandths (.004), place the numeral 4 without the decimal over 1000 ($^{4}/_{1000}$), and so on. Consult the tables in Appendix B for further information on decimal/fraction equivalents.

SUMMARY

• Measurements are very important in the heating and air-conditioning field.

• The two major systems for linear measurements are the English system and the metric system.

• Most measurements in heating and air conditioning use the English system.

• Linear measurement refers to measurements made along a straight line.

• The graduations on a measuring instrument are standard within the United States, and the divisions and lengths are named and specified by definition.

• Accurate dimensions are necessary for the accurate calculation of heating and cooling loads, the provision of proper space for air-conditioning units, and the proper installation of ductwork and piping.

REVIEW QUESTIONS

1. Discuss briefly why the heating and air-conditioning technician needs to understand and be able to take accurate measurements.

2. What U.S. government agency is responsible for maintaining uniform standards of measurement?

3. In your opinion, why is it important that uniform standards of measurement be maintained?

4. Name the two major measuring systems used in the United States.

5. Define linear measurement.

6. Why does the heating and air-conditioning technician need to understand reduced scales?

7. Explain in a paragraph or diagram how a one-inch graduation is subdivided into smaller segments.

8. List and describe three commonly used measuring devices.

9. Figure 3.6 contains a series of lines. Using a twelve-inch scale, determine the length of each line.

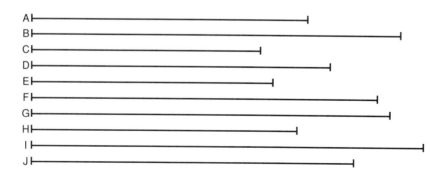

Figure 3.6 Exercise for measuring lines

4

ANGULAR MEASUREMENT

OBJECTIVES

After studying this chapter, you should be able to:

- demonstrate an understanding of angular measurement
- describe the four angles formed at the center of a circle that has been divided into four equal parts
- define and illustrate three other types of angles
- describe the relationship between degrees, minutes, and seconds
- identify angles in a triangle
- determine the third angle in a triangle when the other two angles are known
- explain how there are 180 degrees in a straight line
- read and use a protractor to measure angles

4.1 INTRODUCTION

In addition to understanding linear measurement, the HVAC technician must know how to measure circles and other geometric figures. There are circular spaces in buildings. Ductwork is often circular in shape. Unusual angles are created by the architectural design of many modern buildings. Triangular shapes are created when a room has a cathedral ceiling (Figure 4.1).

Regardless of the shape of the space involved, the HVAC technician must know how to determine areas and volumes in order to determine heating and cooling loads for sizing equipment and for air distribution. (Refer to Appendix A for information on calculating area and volume.)

4.2 ANGULAR MEASUREMENTS IN A CIRCLE

A circle is a closed curve on a plane with all points on the curve the same distance from the center of the circle. A circle contains 360 degrees (Figure 4.2). In other words, a circle can be divided into 360

Figure 4.1 Triangular shape in a building (cross-hatched)

Figure 4.2 Four quadrants of a circle

sections that could be cut from the center of the circle. Each of these 360 sections equals one angular degree. When a circle is divided into four quadrants, each quadrant has 90 degrees.

A useful term in angular measurement is *vertex*. The point where two straight lines intersect is called the vertex (Figure 4.3). Lines extended from the vertex are called *sides*. Spaces between the lines, which taper to a point at the vertex, are called *angles*. These spaces (angles) are measured in degrees. If a vertical line passes through the center of a circle, 180 degrees can be measured between the line to the right of center and the line to the left of center of the circle (see again Figure 4.2).

If a *horizontal* line is drawn through the center of a circle, the portion of the line to the left of center is 180 degrees from the portion of the line to the right of the center of the circle. *A straight line, therefore, contains 180 degrees* (Figure 4.4).

4.3 FOUR KINDS OF ANGLES

Four types of angles are the *right angle, acute angle, obtuse angle,* and *straight angle.*

If a vertical line and a horizontal line cross at the center of a circle as illustrated in Figure 4.5A, the circle is divided into four equal parts. Each of the four equal angles in the circle has 90 degrees (one-fourth of the 360 degrees in the whole circle). Each 90-degree angle is called a right angle.

An angle of less than 90 degrees is called an acute angle (Figure 4.5B). An angle greater than 90 degrees is called an obtuse angle (Figure 4.5C). An angle between two horizontal lines drawn end-to-end through the center of the circle is equal to 180 degrees and is called a straight angle or a straight line (Figure 4.5D).

4.4 ANGLES OF A TRIANGLE

A triangle is a geometric figure with three sides and three inside angles. A triangle can be created by breaking a straight line in two places and then

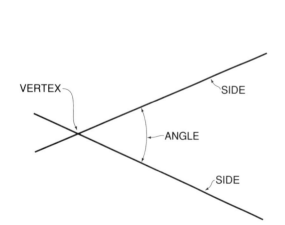

Figure 4.3 Vertex and sides of an angle

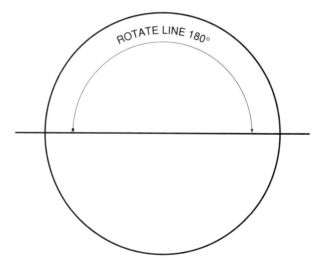

Figure 4.4 A straight angle

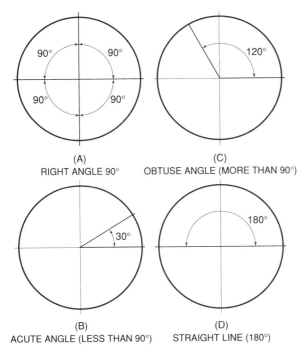

(A)
RIGHT ANGLE 90°

(C)
OBTUSE ANGLE (MORE THAN 90°)

(B)
ACUTE ANGLE (LESS THAN 90°)

(D)
STRAIGHT LINE (180°)

Figure 4.5 Four kinds of angles

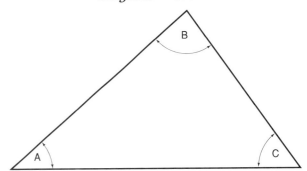

Figure 4.7 The three angles of a triangle

4.5 USING THE PROTRACTOR

The protractor is an instrument consisting of a half-circle with the midpoint (center) marked on the horizontal portion (base) of the protractor. This midpoint, also called the reference point, is clearly marked on the protractor. The half-circle shape of the protractor is divided into degrees, which are usually numbered from right to left and from left to right, so that angles can be measured from the right side or the left side (Figure 4.8).

To use the protractor, place it over the angle to be measured, with the base (horizontal) portion of the protractor directly on the side of the triangle and the point of the angle at the midpoint on the base of the protractor. The angle in Figure 4.9 measures 30 degrees.

With the protractor base directly on the side of the angle and the point of the angle directly under the midpoint, the adjacent side (line) of the triangle intersects the degree scale on the half-circle. The number of degrees of the angle can be read on the degree scale.

If the triangle is too small for the adjacent side to cross the degree scale of the protractor, extend

joining the ends together. The figure that has been created has three sides and three inside angles. Since the figure started as a straight line, the sum of all three angles is equal to 180 degrees (Figure 4.6).

If the sum of two known angles is subtracted from 180 degrees, the difference is the number of degrees in the unknown angle. This rule holds true regardless of the shape of the triangle. In Figure 4.7, angles A + B + C = 180 degrees. *The sum of two enclosed angles subtracted from 180 degrees equals the third angle.* Hence, the equations:

A + B + C = 180 degrees

180 degrees − (A + B) = C

180 degrees − (B + C) = A

180 degrees − (C + A) = B

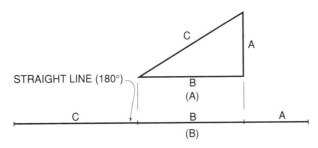

STRAIGHT LINE (180°)

Figure 4.6 How a triangle is formed

Figure 4.8 A protractor (*Photo by Bill Johnson*)

Figure 4.9 Using a protractor

the adjacent side so that the extended line intersects the degree scale.

Special care must be taken in the use of the protractor. The base must be placed directly over the side of the angle being measured, and the point of the angle must be directly under the midpoint of the base of the protractor. The drafter must be accurate in reading the degrees on the half-circle at the point where the adjacent side intersects. For accuracy in measuring angles, it is very important for all of these precautions to be taken and for the protractor to be held firmly in place without letting it slip.

4.6 DEGREES, MINUTES, AND SECONDS

Sometimes a measurement requires more accuracy than can be obtained by measuring whole degrees. To allow this accuracy, each degree can be subdivided into sixty sections called *minutes*. Each minute can be further subdivided into sixty sections called *seconds*. (Remember that minutes and seconds as they refer to angular measurement are different from minutes and seconds as a measurement of time.) Minutes and seconds divide parts of a circle into very small segments. For example, in a right angle (90 degrees), there are 5400 minutes (90 degrees x 60 minutes per degree). There are 324,000 seconds in a right angle (90 degrees x 60 minutes x 60 seconds per degree).

A small variation in an angular measurement makes a great difference. This difference is especially obvious in navigation for air, sea, and space travel. For example, the distance across the United States is approximately 3000 miles. If a navigator

makes a one-degree mistake in charting the destination across the country, the vehicle will miss the mark by more than 52 miles!

It should be noted that computers and advanced technology can be used to figure angular measurements to extremely minute (very small) numbers. Space technology and high tech equipment require measurements smaller than the human eye can discern. Extremely small measurements are rarely needed by the HVAC technician, however, who relies on instruments like the protractor to take measurements in whole degrees.

SUMMARY

- A circle is a closed curve on a plane with all points on the curve the same distance from the center of the circle.

- The vertex is the point where two intersecting lines meet.

- The lines extending out from the vertex are called sides.

- The space between the lines extending out from the vertex is the angle between those lines.

- A straight line (or straight angle) contains 180 degrees.

- A right angle contains 90 degrees.

- An acute angle contains fewer than 90 degrees.

- An obtuse angle contains more than 90 degrees.

- A triangle has three sides and three inside angles.

- A protractor is an instrument used for measuring angles.

- Angles can be subdivided into tiny segments called minutes and seconds.

REVIEW QUESTIONS

1. Why must the heating and air-conditioning technician know how to measure circles and other geometric figures in his job?

2. List and define the following terms as they refer to circles:
   A. vertex
   B. sides
   C. angle
   D. straight angle
   E. right angle
   F. acute angle
   G. obtuse angle

3. Define a circle. How many degrees are contained in a circle?

4. Define a triangle. How can a straight line be changed into a triangle?

5. If two of the angles of a triangle are known, how can the number of degrees in the third angle be determined?

6. Define a protractor. What is meant by the reference point of a protractor?

7. Why is it essential that the drafter be very careful in using the protractor?

8. How many minutes are contained in one degree?

9. How many seconds are contained in one minute?

10. In your opinion, why is it usually not necessary for the heating and air-conditioning technician to measure angles in minutes and seconds?

ADDITIONAL STUDENT EXERCISES: MEASURING ANGLES

1. Using the protractor, measure the two angles shown in Figure 4.10 and record your answers.

Answers: Angle A = _____ degrees
            Angle B = _____ degrees

2. Using the protractor, determine all the angles in each triangle in Figure 4.11. Record your answers.

Answers: A = _____ degrees   D = _____ degrees
            B = _____ degrees   E = _____ degrees
            C = _____ degrees   F = _____ degrees

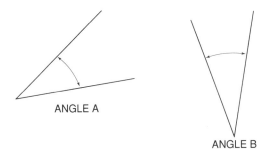

ANGLE A

ANGLE B

Figure 4.10 Exercise 1

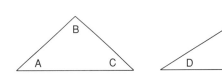

Figure 4.11 Exercise 2

5

LEARNING TO USE THE ARCHITECT'S SCALE

OBJECTIVES

After studying this chapter, you should be able to:

- describe several sizes and kinds of architect's scale rules

- demonstrate the ability to use a one-inch scale

- explain why a drafter needs to be able to read and use an architect's scale

- state several criteria used by a drafter in order to select the proper scale for a drawing

- demonstrate the ability to use the $1/8$-, $1/4$-, $1/2$-, and $3/4$-inch scales

NOTE: Refer to an architect's scale while studying this chapter.

5.1 INTRODUCTION

In drafting and freehand sketching, it is important for the entire drawing to be in proportion. With freehand sketching, drawings are made to *near scale*. In the drafting room, drawings are drawn *to scale*. Drawing a building to *full scale* would be unthinkable, because the finished drawing would be the same size as the building. Even if plans could be drawn and printed to full scale, they would be too bulky for use on the construction site. To keep drawings in proportion and plans a practical size, architects, engineers, and drafters depend on scale rules (often called simply *scales*)

for measurement. With these scales, drawings can be reduced to a size suitable for the construction site or the shop. Also, since the drawings are *to scale*, (even though they are not *to full scale*), all lines on the plan are in proportion. Choosing the scale is important because the architect, engineer, or drafter tries to draw plans as large as possible within the confines of the page (the larger the plan, the easier it is to read). There are many reduced scales on the architect's scale rule. This chapter will discuss the ones that are most commonly used.

5.2 THE ARCHITECT'S SCALE RULE

Of the various types of scale rules in the drafting room, the architect's scale rule is probably used most often. It is also frequently carried on the job in the air-conditioning field. The best known is a triangular scale rule twelve inches long containing a total of eleven scales (Figure 5.1).

Another version of the triangular scale rule is only six inches long. This six-inch scale rule also contains eleven scales.

A flat scale rule six inches long is very popular because it contains eight of the most commonly used scales and it fits easily into the shirt pocket (Figure 5.2). Manufacturers sometimes give these flexible, plastic, six-inch scale rules with advertisements printed on them to architects, engineers, and

Figure 5.1 Twelve-inch architect's scale (*Photo by Bill Johnson*)

drafters. Some architects, engineers, and drafters prefer twelve-inch, flat scale rules which have the same scales as the six-inch versions.

The architect makes a note on the blueprint telling which scale was selected for that particular drawing.

5.3 UNDERSTANDING THE ARCHITECT'S SCALE RULE

At first glance, the architect's scale rule appears complicated and confusing, because so many different scales are shown on the same measuring device. On the twelve-inch, triangular rule, two sides of the rule hold four different scales each, and the other side holds three additional scales. Some of these scales are read from left to right, and others are read from right to left. The size of each scale is indicated at the end of the rule.

The scale size of a drawing is actually a ratio of the size of the drawing compared to the size of the building (or other object being drawn to scale). A scale of 1/4" = 1'-0" means that each one-fourth inch on the drawing represents one foot on the building itself. Other frequently used scales are as follows:

1/8" = 1'-0"

1/2" = 1'-0"

3/4" = 1'-0"

1" = 1'-0"

5.4 PRACTICING WITH THE ONE-INCH SCALE

To learn how to read and use the architect's scale rule, you will need the following items: a triangular twelve-inch architectural scale rule, a piece of paper, and a pencil. (A flat architectural scale rule could be used if the triangular rule is not available.) Then follow these step-by-step directions:

1. Place the architectural scale rule in front of you so that the "1" on the rule is on your left and the "1/2" is on your right. The numeral "1" stands for the one-inch scale, meaning that each inch represents one foot on the original building (or other object). You read the one-inch scale from *left to right*, using the markings from "1" to "10" to represent measurements for the 1" = 1'-0" scale.

2. Next, look at the "1/2" at the right hand end of the scale. Reading (backward) from *right to left*, notice that the 1/2" = 1'-0" scale has small markings from "2" to "20." On this scale, each one-half inch represents one foot of the original building or object. You will use the "1/2" scale later.

Figure 5.2 Six-inch architect's scale (*Photo by Bill Johnson*)

Figure 5.3 One-inch scale on the architect's scale (*Photo by Bill Johnson*)

3. Now concentrate on the one-inch long section located at the left of the scale between the "1" and the "0" (Figure 5.3). Next to the "1" signifying the one-inch scale are tiny subdivisions and the numerals "9," "6," "3," and "0." This section of the rule is used for lines that represent less than one foot in measurement. Each numeral within the small section represents that many inches using the one-inch scale; that is, the "9" stands for a nine-inch measurement, the "6" stands for six inches, and so on. By counting the tiny marks between the numbers, you can also figure unlabelled inches, half-inches, and quarter-inches on this scale.

4. Now you are ready to draw lines using the 1" = 1'-0" scale. To draw a line representing a measurement of six feet, for example, count past the 1, 2, 3, and so on to find the numeral 6 on the "1" scale. Then draw a line between the "6" and the "0," being careful to drag the pencil and not tear the paper. (Usually, a right-handed drafter will move the pencil from left to right, and a left-handed drafter will move from right to left.) The line you have drawn should be six inches long, representing six feet on a scale of 1" = 1'-0".

To draw a line standing for six feet and three inches (6¹/4 feet), use the same scale to draw the line from the "6" to the "0" and continue past the "0" to the small "3" (which represents three inches more). The line represents six feet and three inches (6¹/4 feet) on a scale of 1" = 1'-0".

5.5 USING THE ONE-HALF INCH SCALE

If you understand the "1" scale as described above, it should not be difficult to read the "¹/2" scale too. To read the ¹/2-inch scale, use the same face of the triangular scale, but this time you read from *right to left* (Figure 5.4). Again, follow the step-by-step procedure:

1. Look at the right hand end of the architectural scale rule at the fraction "¹/2 ," denoting the ¹/2" = 1'-0" scale. Each one-half inch on this scale represents one foot on the actual building or object.

2. Notice the marks just to the left of the "¹/2" designation. This small section shows one foot (12 inches) broken down into inches and fractions of

Figure 5.4 One-half inch scale on the architect's scale (*Photo by Bill Johnson*)

an inch. In Figure 5.4, the section up to the "0" has 24 tiny lines so that you can accurately draw measurements containing inches or fractions of an inch. Each of the smallest division lines represents one-half inch. The longer divisions represent whole inches, and the longest divisions represent three inches each.

3. On your twelve-inch architectural rule, read *from right to left* every other number from "2" (standing for two feet) to "20" (standing for twenty feet). The line you draw will extend between the "0" at the right hand end of the scale and the designated number. (For this step, Figure 5.4 is not adequate, since it shows only a small segment of the rule.)

4. Now you are going to use the "1/2" scale to draw a line representing twelve feet. Start at the "0" on the right hand end of the scale and read backward by two's to the numeral "12." Draw the line between the "12" and the "0." Your line is six inches long, and it represents twelve feet when the 1/2" = 1'-0" scale is used.

5. To draw a line representing twelve feet and nine inches (12 3/4 feet), draw the line from the numeral "12" to the "0" just as you did before, and then continue on for three major subdivisions to the right of the "0," representing nine inches more than the twelve feet.

5.6   USING OTHER SCALES

The other scales function in the same way as the ones you have just used: the small area subdivided at each end of the rule ending with a "0" represents fractional parts of one foot, with the number of subdivision lines indicating what fraction is represented. The small area always represents 12 inches, and each mark inside may stand for one inch, one-half inch, or other fractional part, depending on what scale you are working with.

You can use other scales accurately by referring to the correct position on the triangular architectural scale. You must be careful to read the numbers from the correct end of the scale, since two different scales occupy the same space on the face of the triangular scale, and the numbers for the different scales overlap. The scale on the left reads *left to right*, and the scale on the right reads *right to left*. Figure 5.5 shows

measurements that have been made with the 3/8" scale, the 1/8" scale, and the 1/2" scale.

5.7   THE MOST COMMONLY USED SCALES ON THE ARCHITECT'S SCALE RULE

The drawing paper or drawing film used by architects and engineers is usually 24" x 36" or 36" x 42". The 24" x 36" drawing paper or film is almost always used for drawing plans for residential and small commercial buildings. If the building is 100 feet by 200 feet, the architect and engineer can put the entire floor plan on a 24" x 36" plan with a scale of 1/8" = 1'-0". (Figure 5.4 contains the 1/8" = 1'-0" scale along with the 1/2" = 1'-0" scale, but it can best be read by inverting the page.) A building of 20 feet x 50 feet can be shown easily on a 24" x 36" sheet when the 1/4" = 1'-0" scale is selected.

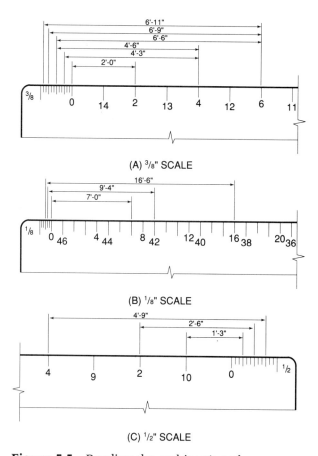

Figure 5.5   Reading the architect's scale

For larger buildings, the architect usually chooses the $^1/8$" = 1'-0" scale. In some cases smaller scales are used, but the plans become very difficult to read with any accuracy.

Depending on the size of the building and the size of the drawing paper (or drawing film), the architect and engineer usually draw floor plans of buildings using the $^1/8$" = 1'-0" or the $^1/4$" = 1'-0" scales. Details, diagrams, and sections are sometimes drawn to larger scales such as $^3/4$" scale, 1" scale, and even $1^1/2$" scale.

SUMMARY

- It would not be practical to draw building plans to full scale.

- The twelve-inch, three-sided architect's scale rule has eleven scales.

- The "0" on each scale separates the inches from the feet on the architect's scale.

- There are several kinds and types of architect's scale rules; however, the standard scales should be the same on all rules.

- The smaller the scale, the smaller the drawing.

- The most commonly used scales on the architect's scale are the $^1/8$" scale and the $^1/4$" scale.

- Details, diagrams, and sections are usually drawn to a larger scale such as $^3/4$", 1", and $1^1/2$" scales.

REVIEW QUESTIONS

1. Name all the scales found on the twelve-inch, three-sided architect's scale.

2. What are the two most commonly used scales on the architect's scale rule?

3. Discuss the factors considered by an architect or engineer when choosing a scale for drawing a floor plan on a sheet.

4. With an architect's scale rule, measure each line in Figure 5.6, using the $\frac{1}{8}$", $\frac{1}{4}$", $\frac{1}{2}$", $\frac{3}{4}$", and the 1-inch scales for measuring each line. (You will have five different answers for each line measured, with a different answer for each scale used.)

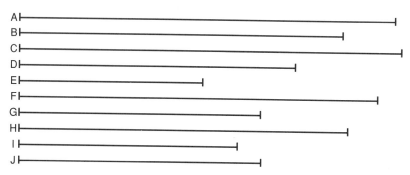

Figure 5.6 Reading the architect's scale

ADDITIONAL STUDENT EXERCISE: USING THE ARCHITECT'S SCALE

Using the $\frac{1}{8}$" scale and the $\frac{3}{8}$" scale, give the indicated dimensions for Figure 5.7.

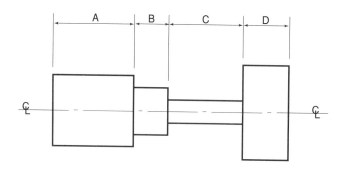

USING THE ARCHITECTURAL SCALE, GIVE THE INDICATED DIMENSIONS USING THE $\frac{1}{8}$ SCALE AND THE $\frac{3}{8}$ SCALE.

SCALE	DIMENSIONS			
ANSWER	A	B	C	D
$\frac{1}{8}$ SCALE				
$\frac{3}{8}$ SCALE				

Figure 5.7 Exercise 1

6

LEARNING TO USE THE ENGINEER'S SCALE

OBJECTIVES

After studying this chapter, you should be able to:

- discuss the purpose of using the engineer's scale rule

- give examples of types of plans for which the engineer's scale rule is commonly used

- describe three different kinds of engineer's scale rules

- explain and demonstrate how the 10 scale is used

- list other commonly used scales and demonstrate how to use them

NOTE: Refer to an engineer's scale while studying this chapter.

6.1 INTRODUCTION

The engineer's scale rule and the architect's scale rule are very similar. The purpose of the engineer's scale rule is to reduce the size of a plan, to scale, so that the plan can be put on a workable-size drawing paper or drawing film. The civil engineer often uses the engineer's scale rule because it adapts easily to making drawings of large tracts of land that are to be developed.

With the engineering scale rule, almost any reduced scale can be used so long as it relates to tens, hundreds, thousands, and so forth. Site plans and plot plans are usually drawn using the engineer's scale. HVAC technicians use site plans and plot plans to locate air-conditioning piping and equipment on the site. For this reason, technicians must know how to use the engineer's scale rule.

On smaller jobs, architects often use the engineer's scale to develop the site plan and show the grading plan for the site. If the site is small, a smaller scale on the *architect's rule* (like $1/16$" = 1'-0" or even $1/32$" = 1'-0") may be selected. In most cases, however, the architect prefers the engineer's scale for developing small scale drawings.

6.2 THE ENGINEER'S SCALE RULE

The engineer's scale rule is so designed that large dimensions can be reduced to a small scale. There are several types of engineer's scale rules. The most common type is the twelve-inch, triangular scale rule (Figure 6.1). Each of the three sides has two scales. The triangular engineer's scale is also available in a six-inch model. Some drafters find this six-inch rule easier to use than the twelve-inch version.

Some engineers and drafters prefer to use a twelve-inch, flat (or two-sided) engineer's scale rule. These flat rules have only two scales on each side; however, these four scales are the most com-

Figure 6.1 Twelve-inch engineer's scale *(Photo by Bill Johnson)*

monly used scales in the engineering field. The flat rule is also available in a six-inch model which is popular because it is easily carried in the shirt pocket (Figure 6.2). Like the flat twelve-inch rule, it is limited to two scales on each side rather than two scales on each of the three sides of the triangular rule.

Equipment manufacturers and other retailers and businesses distribute flexible plastic engineer's scale rules with company names on them for advertisement. These rules are similar to the flat rule described above. They are popular because they are easily carried and they are usually obtained for free.

6.3   WHAT MAKES THE ENGINEER'S SCALE UNIQUE

The architect's scale is divided into units of 2, 4, 6, 9, 12, and so forth, which relate to feet and inches very readily. The engineer's scale is divided into units of 10, 20, 30, 40, 50, and 60. One must be careful not to mix the engineer's scale with the architect's scale. Even on the 10 scale where each one-inch segment is numbered, the inch segment on the engineer's rule is divided into tenths (with ten division marks per inch). Figure 6.3 shows the divisions on a partial 10 scale. On a standard foot rule, each inch is separated into eight divisions, signifying halves, quarters, and eighths of an inch.

6.4   USING THE ENGINEER'S SCALE RULE

The triangular, twelve-inch engineer's scale has six scales (two on each side). The scales are numbered 10, 20, 30, 40, 50, and 60. Each scale is then subdivided so that each inch has ten major subdivisions on the 10 scale, twenty major subdivisions on the 20 scale, thirty major subdivisions on the 30 scale, and so on for the other three scales. Some scales have additional subdivisions to make the rule easier to use. The person who understands how to use one scale can use the other scales because the procedure is the same.

The 10 scale is often used by architects and engineers. The scale is identified by the number 10 at the left end of the rule, and inches are indicated by numbers reading left to right (Figure 6.3).

Each inch is divided into ten equal parts. On this scale, one inch could equal ten feet (usually stated as 1" = 10'-0"), and each subdivision between the inch marks would equal one foot. The scale could also be used so that one inch equals 100 feet, with each subdivision between the inch marks representing ten feet. A scale of one inch equals 1,000 feet means that each subdivision between the inch marks equals 100 feet. The scale could even be used for one inch equals 10,000 feet or 100,000 feet. The principle for reading and using the scale is the same regardless of the length represented.

The 30 through 60 scales are set up the same way (Figures 6.4 and 6.5). Although numbering is not indicated for each inch on these scales, reading the scales is not difficult. For example, the 30 scale has major divisions numbered 6, 8, 10, 12, and so forth up to 36. The first inch could represent 30 feet, even though the numbering on the scale is 3. The first inch is halfway between the 2 and 4 (or 3, even if it is not numbered). This scale can be used to represent 1 inch equals 30 feet, 300 feet, 3,000 feet, and so on.

Figure 6.2 Six-inch engineer's scale *(Photo by Bill Johnson)*

Figure 6.3   10 SCALE on the engineer's rule *(Photo by Bill Johnson)*

Figure 6.4   30 SCALE on the engineer's rule *(Photo by Bill Johnson)*

Figure 6.5   60 SCALE on the engineer's rule *(Photo by Bill Johnson)*

6.5   COMMONLY USED SCALES ON THE ENGINEER'S SCALE RULE

The most commonly used scales on the engineer's scale are the 10, 20, and 30 scales. The architect and the engineer can usually put all the information needed by the HVAC technician on a site plan drawn to a 1" = 10'-0", 1" = 20'-0", or 1" = 30'-0" scale. Site plans that are used by HVAC technicians are usually localized to the building rather than to large tracts of land. Figure 6.6 illustrates the way the 10 scale can be used for many different units of measurement.

SUMMARY

• It is not practical to draw large tracts of land to a large scale. It is hard to get such drawings to fit on standard sheets of drawing paper or drawing film.

• Architects often use the engineer's scale to draw site plans.

• The twelve-inch, triangular engineer's scale rule has a total of six scales.

• Each scale on the engineer's scale rule can be used for at least four scales: 10s, 100s, 1,000s, and 10,000s.

• There are several types of engineer's scale rules, but the scales are the same. Some of the scales described in this chapter do not appear on all engineer's scale rules.

• The most commonly used scales on the engineer's scale are the 10, 20, and 30 scales.

• The larger the scale number, the smaller the drawing will be.

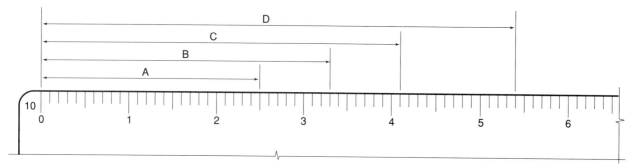

LINE	SCALE	DISTANCE
A	1" = 10'	25'
B	1" = 10'	33'
C	1" = 10'	41'
D	1" = 10'	54'
A	1" = 100'	250'
B	1" = 100'	330'
C	1" = 100'	410'
D	1" = 100'	540'

LINE	SCALE	DISTANCE
A	1" = 1000'	2500'
B	1" = 1000'	3300'
C	1" = 1000'	4100'
D	1" = 1000'	5400'
A	1" = 10,000'	25,000'
B	1" = 10,000'	33,000'
C	1" = 10,000'	41,000'
D	1" = 10,000'	54,000'

Figure 6.6   Reading the engineer's scale

REVIEW QUESTIONS

1. Explain when and why an architect uses an engineer's scale rule.

2. Explain when and why HVAC technicians use the engineer's scale rule.

3. What scales on the engineer's scale rule are used most often? Why?

4. Can the architect's scale and the engineer's scale be interchanged on the same drawing? Explain.

5. Using the engineer's scale, if one inch equals one hundred feet, each subdivision of the inch represents _____ feet.

6. Using the engineer's scale, if one inch equals 1000 feet, each subdivision of the inch represents _____ feet.

ADDITIONAL STUDENT EXERCISE: USING THE ENGINEER'S SCALE

1. Using the indicated scales in Figure 6.7, determine the distance of each line. Record your answers in the space provided.

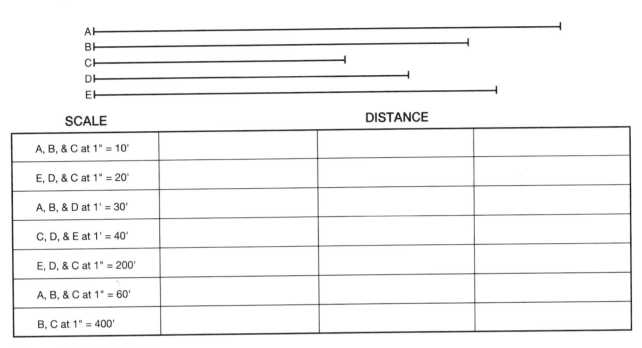

SCALE	DISTANCE		
A, B, & C at 1" = 10'			
E, D, & C at 1" = 20'			
A, B, & D at 1' = 30'			
C, D, & E at 1' = 40'			
E, D, & C at 1" = 200'			
A, B, & C at 1" = 60'			
B, C at 1" = 400'			

Figure 6.7   EXERCISE 1

2. Using the engineer's scale, measure the property lines and total the overall perimeter of Figure 6.8. Record your answers in the space provided.

Figure 6.8   EXERCISE 2

MEASURING A PLOT OF LAND

	LINE A-B	LINE B-C	LINE C-D	LINE D-E	LINE E-F	LINE F-G	LINE G-A
1" = 10'-0" SCALE							
1" = 20'-0" SCALE							
1" = 30'-0" SCALE							

7

SYMBOLS AND ABBREVIATIONS

OBJECTIVES

After studying this chapter, you should be able to:

- define and give examples of abbreviations
- define and give examples of symbols
- define and give examples of acronyms
- discuss the use of nonstandard symbols and abbreviations on drawings
- discuss the advantages of using a Computer-Assisted Design (CAD) library of symbols
- explain the purpose of the "Legend and Symbol Schedule"
- find examples of abbreviations and symbols used together

7.1 INTRODUCTION

In blueprint reading, abbreviations and symbols are used for similar reasons— to save time for the drafter and to save space on the plan. Abbreviations are key letters of words that are used to denote the complete word. When abbreviations are used to replace whole words on a drawing, the drawing is less cluttered and easier to read. Symbols are graphic representations of the building materials and components required to complete a building. Using symbols on a drawing allows the drafter to concretely and succinctly communicate the infor-

mation needed by the person reading the plans. The drafter and the person reading the drawings are responsible for becoming familiar with standard abbreviations and symbols in order to communicate clearly with each other.

7.2 STANDARDIZED ABBREVIATIONS AND SYMBOLS

Standardized abbreviations and symbols for the various trades have been developed by organizations like the American National Standards Institute (ANSI) and the American Society of Heating, Refrigeration, and Air Conditioning Engineers (ASHRAE). These standard abbreviations and symbols are generally used by architects and engineers; however, architects and engineers sometimes create their own symbols and abbreviations to represent materials and equipment on drawings. If symbols and abbreviations are not standard, they are usually noted on the drawing.

Some abbreviations for materials, equipment, and titles are called *acronyms*. An acronym is an abbreviation formed by using the first letter of each word. Some examples of acronyms are CFM (cubic feet per minute), FPM (feet per minute), BTU (British thermal unit), and GPM (gallons per minute). ASHRAE is an acronym for the American Society of Heating, Refrigeration and Air Conditioning Engineers.

7.3 VARIATIONS IN ABBREVIATIONS

Some single letter abbreviations are used to represent several different words. The letter "R" stands for *radius, Rankine, road, room,* and *thermal resistance.* Generally speaking, the location of the single letter abbreviation indicates what that particular abbreviation stands for.

Some words have more than one abbreviation. The word *DOWN* can be written as *DWN* or *DN.* Either abbreviation is correct, but the drafter should choose one and then use it throughout the drawing.

Some abbreviations (single letter or several letters) have a period after the last letter to eliminate confusion for the reader. It is not necessary to place a period after an acronym.

7.4 VARIATIONS IN SYMBOLS

When symbols are drawn by hand, templates are usually used (Figure 7.1). Templates are commercially prepared patterns made of sheets of plastic with the shapes of various symbols cut out of the plastic. To insert a symbol at the proper place on the drawing, the drafter draws around the cut-out portion of the template. If the symbol is a complicated one, two or more templates may be required.

Computer-aided design (CAD) systems have a library of symbols from which the exact symbol can be selected and plotted on the drawing. This library contains symbols for various materials and pieces of equipment in each section of drawings (architectural, plumbing, mechanical, and electrical sections).

In some cases, there is no standard symbol for a material or piece of equipment. For example, there is no standard symbol for *EARTH* when shown on the plan view of a drawing. When there is no standard symbol, the drafter notes on the drawing what the material or equipment is.

Some symbols for the same material are shown differently on different drawings. For example, when brick is shown on the elevation views of a building, the drafter usually shows the brickwork as it will look when the wall is complete. The mortar joints and the brick are shown in true perspective. When

(A) GENERAL USE TEMPLATE

(B) PLUMBING FIXTURE TEMPLATE

Figure 7.1 Typical templates

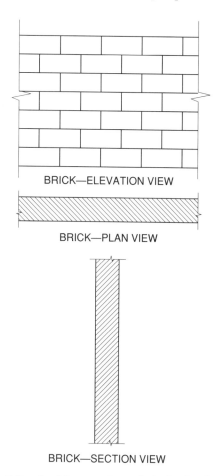

BRICK—ELEVATION VIEW

BRICK—PLAN VIEW

BRICK—SECTION VIEW

Figure 7.2 Architectural symbols for brick

drawn on the floor plan, this brick wall is shown with crosshatching (parallel lines drawn close together at a 45-degree angle). It is important for the reader of the drawings to keep these variations in mind (Figure 7.2).

7.5　LEGEND AND SYMBOL SCHEDULES

In most cases, the architect and engineer include a listing of symbols used on the drawings called the "Legend and Symbol Schedule." It is usually placed on the first drawing of each section of the set of drawings (architectural, mechanical, plumbing, and electrical). When this schedule is present, each symbol used in that section is shown on the schedule whether it is a standard symbol or one created by the drafter.

7.6　ABBREVIATIONS AND SYMBOLS USED TOGETHER

Abbreviations and symbols are combined on working drawings to relate information to the builder. Abbreviations and symbols are used jointly to describe the sidewall supply register in Figure 7.3. The arrow indicates the discharge of supply air from the register. The air quantity is shown on the arrow with the abbreviation CFM (cubic feet per

Ⓐ SYMBOL INDICATES SIDE WALL REGISTER. THIS REGISTER WILL BE SHOWN ON THE "REGISTER, GRILLE, AND DIFFUSER SCHEDULE" ON THE PLANS AND WILL BE DESCRIBED OR SPECIFIED IN THIS SCHEDULE

Figure 7.3　Use of symbols and abbreviations

minute). A circle, square, or other symbol is placed at the arrowhead to indicate which register to look for in the Register and Grille Schedule. This kind of combination of symbols and abbreviations is often found on mechanical plans.

7.7　EXAMPLES OF ABBREVIATIONS AND SYMBOLS

Some of the standard abbreviations and symbols used in blueprint reading are listed in the appendices at the end of this book. Appendix C contains standard abbreviations, and Appendix D contains standard symbols. The student is advised to study both appendices now and to refer to them later as needed. The construction drawings folded in the back cover demonstrate the use of abbreviations and symbols on actual construction drawings.

SUMMARY

- Abbreviations and acronyms are used on drawings to relate information in a shortened, easy-to-read form.

- Standard abbreviations have been developed for words that are used often in the construction field.

- Some single letter abbreviations are used to represent several different words.

- If a word has more than one abbreviation, the same abbreviation should be used throughout a set of drawings.

- Symbols and abbreviations are often used together to relay information to the reader.

- Symbols are graphic representations of the building materials and components that are shown on construction drawings.

- Templates are used as patterns for manually drawn symbols.

- Computer-assisted design (CAD) systems have a library of standard symbols to choose from.

- A symbol schedule on the drawings explains what each symbol stands for. This schedule is especially helpful if nonstandard symbols have been used.

REVIEW QUESTIONS

1. Why are abbreviations and acronyms used on drawings?

2. Define *symbols*. Why are symbols used on drawings?

3. What is a template?

4. Why do architects and engineers sometimes use non-standard symbols?

5. What information is placed on the "Legend and Symbol Schedule"?

6. Using Appendix C at the end of the textbook, give the abbreviations for the following terms:

   A. Basement
   B. Ceiling
   C. Cross section
   D. Fire extinguisher
   E. Glass
   F. Men's rest room
   G. Pounds per cubic foot
   H. Thermometer
   I. Air conditioning
   J. Bathtub
   K. Cement floor
   L. Cold water
   M. Drinking fountain
   N. Fireproof
   O. Hot water heater
   P. Shut-off valve

   Q. Velocity
   R. Access panel
   S. British thermal unit
   T. Conduit
   U. Fabricate
   V. Gallons per minute
   W. Elevator
   X. Pressure-reducing valve
   Y. Vent
   Z. Alternating current
   AA. Breaker
   BB. Current
   CC. Direct current
   DD. Solder
   EE. Volt

7. The following abbreviations are familiar to architects and engineers. Explain what each one stands for.

   A. ARCH
   B. AGA
   C. ASME
   D. BS
   E. NFPA

   F. UL
   G. ASHRAE
   H. ASRE
   I. AIEE

ADDITIONAL STUDENT EXERCISES: FREEHAND DRAWING OF SYMBOLS

DIRECTIONS: Draw freehand the following symbols that can be found in Appendix D.

1. Gate valve
2. Pressure-reducing valve
3. Pipe union
4. Backdraft damper
5. Pipe turning down
6. Butterfly valve
7. Angle gate valve
8. Compressed air pipe
9. Fuel gas pipe
10. Floor drain
11. Hot water heater
12. Pipe sleeve on a pipe
13. Three-way wall switch
14. Electrical lighting panel (distribution panel)
15. Power panel (electrical service panel)
16. Electrical motor
17. Electrical isolating switch (circuit breaker)
18. Turning vanes in a round-throated elbow
19. Electric heater in ductwork
20. Plan view, rectangular duct turned up
21. Flexible duct connection
22. Plan view, stud partition wall
23. Plan view, wall with concrete block and brick
24. Finished grade line, as shown on site plan
25. Deflector, behind a register or grille

8

HOW WORKING DRAWINGS ARE CREATED

OBJECTIVES

After studying this chapter, you should be able to:

- define working drawings
- list tools and equipment used to create working drawings
- define CAD
- describe the process by which working drawings are manually created
- describe the process by which working drawings are created with a computer
- list advantages and disadvantages of using a computer to create working drawings

8.1 INTRODUCTION

When an owner decides to construct a building, the following procedures are followed: (1) financing is arranged as necessary for the construction; (2) an architect and engineer are hired to design the building; and (3) discussions take place with the architect and engineer so that the building as designed will meet the needs of the owner. The plans resulting from this collaboration are called working drawings.

8.2 WHAT ARE WORKING DRAWINGS?

Working drawings are a set of plans containing graphic information and written instructions for a construction project. These plans give dimensions and details as to how different materials and construction members are to be joined in the construction. Working drawings are the *original drawings*.

Working drawings may be drawn by hand or with the aid of a computer. Until recent years all working drawings had to be drawn manually. Today many companies use computers to create working drawings.

Working drawings are drawn on translucent drawing paper that is too fragile for use on the construction site because the paper will not hold up under heavy use by construction workers. For this reason, copies of working drawings are made to use on the construction site as the building is being constructed. These copies are called "blueprints" or "construction drawings." How blueprints (construction drawings) are made from working drawings will be discussed in Chapter 11.

8.3 HOW ARE WORKING DRAWINGS MANUALLY CREATED?

Hand-drawn working drawings are drawn to scale using T-squares, parallel bars, or a drafting machine

(Figure 8.1). T-squares and parallel bars are used to draw horizontal lines only. Vertical and angular lines are drawn using triangles in conjunction with T-squares and parallel bars.

Instead of using T-squares and parallel bars, the drafter may choose to use a drafting machine. When a drafting machine is used, the rotating head can be adjusted to draw horizontal, vertical, and angular lines without the use of triangles.

Compasses and other drafting instruments are used to draw circles and arcs. Dividers and scale

rules are used to measure distances and lengths (Figure 8.2). Review Chapter 5 for information on reading the architect's scale, and Chapter 6 for information on reading the engineer's scale. The use of the protractor was discussed in Chapter 4.

Working drawings are made on translucent drawing film using a pencil or pen and ink. The plan size can range from about 8$\frac{1}{2}$" x 11" to 36" x 42". Some plans are even larger; however, a plan much larger than 36" x 42" is hard to handle on the job. The most common size for working drawings in residential and small commercial work is 24" x 36".

8.4 HOW ARE WORKING DRAWINGS CREATED WITH A COMPUTER?

Drafters and designers who have had special training often use computers to create working drawings. This method is called "CAD" (computer-aided design). Computer-aided design is also called computer-aided drafting. The CAD operator draws segments of drawings on the computer monitor (screen) and puts these segments together to produce a final drawing.

The advantages of computer-aided design include accuracy, consistency, and speed. Another major advantage is the ease of making changes. Line weights, lettering, and symbols are consistent and neat. Using computer-aided design allows the designer to create several design alternatives at the same time, since the repetitive aspects of designing can be handled easily and quickly. In most cases, a well-trained CAD operator has more time for creative efforts and less time expended on the actual preparation of drawings.

The computer drafting machine is composed of a software program for computer-assisted drafting and hardware consisting of a keyboard and related input equipment (mouse), a processing unit, and a monitor (Figure 8.3, page 42). The processing unit receives information from the input devices (keyboard with function keys and the mouse). Memory is built into the processing unit to store information. Information is also stored on computer disks (Figure 8.4, page 42).

There are two major sizes of disks. One is a soft disk, called the "floppy disk," measuring about 5$\frac{1}{4}$" in diameter and consisting of a round disk enclosed within a jacket (Figure 8.4B). An opening in the

Figure 8.1 Manual drafting equipment

jacket enables the computer to store information on the disk inside. The other disk, 3$\frac{1}{2}$" in diameter, is more durable, and stores more information than the 5$\frac{1}{4}$" type. The 3$\frac{1}{2}$" disk is encased in plastic with a metal window-like device to protect the disk when it is not in the computer (Figure 8.4A). When the disk is inserted into the computer, this window is opened, and the computer records information onto the disk.

When a plan has been completed on the computer, it is printed on translucent paper or drawing film with a printer or plotter. The printer draws on the translucent paper or film with ink drafting pens. The printer is capable of drawing with several line weights and many colors if desired. The printer prints the information received from the computer. Paper is held in place in the printer, and pens are guided across the paper by the mechanism to com-

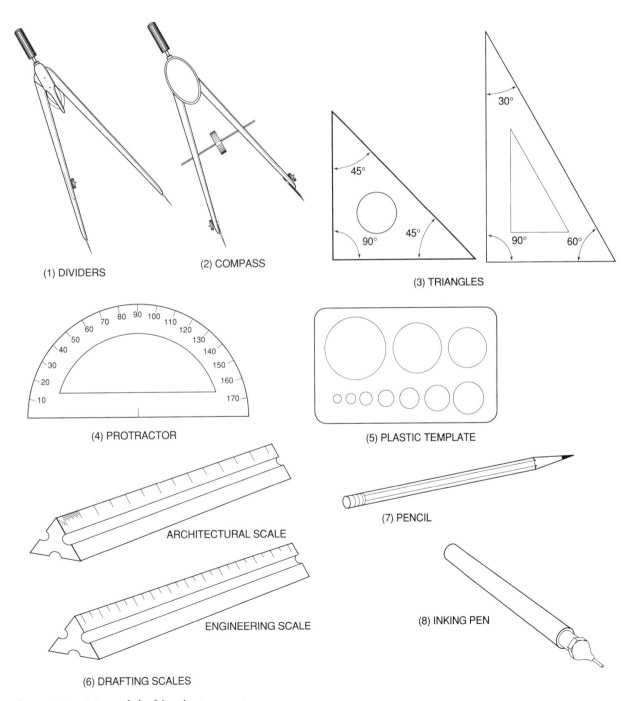

Figure 8.2   Manual drafting instruments

plete the drawing. The pens are moved horizontally across the paper, and the paper is moved vertically.

Figure 8.3 The computer drafting machine

(A) 3¹/₂" DISK

(B) 5¹/₄" DISK

Figure 8.4 3¹/₂" and 5¹/₄" computer disks

With the two movements geared together, circles and other curved lines can be drawn (Figure 8.5).

When CAD machines first came on the market, they were so expensive that only large companies could afford them. They are still expensive; however, the price has dropped enough that more companies are able to buy and use them.

Major advantages of having working drawings drawn by a CAD system are that the line weights are more uniform, the symbols are always the same, and the lettering is precise and easy to read. Another advantage is that when drafting is done by computer, alternate designs can be created quickly with a minimum amount of effort. Some companies use a system whereby color keys designate which drawings were done by the various designers (architectural, mechanical, electrical, plumbing, etc.). By overlaying the various colors of sheets of drawings, better coordination between departments is assured. If two or more items have been designed by different departments to occupy the same space in the building, potential problems are apparent and can be corrected. The quality of drawings done manually depends on the technique and neatness of the drafter. If the drafter is inexperienced or careless, lettering may not be uniform and the quality of drafting may be poor.

Figure 8.5 Pen plotter (printer) machine

8.5 DRAWING PAPER OR FILM USED FOR WORKING DRAWINGS

Several drawing materials are used for working drawings. One expensive drawing sheet is a high grade of linen. This linen is tightly woven and treated with a film that makes the linen smooth and translucent like a drawing paper. The finish of the linen has a texture that allows a graphite pencil and drawing ink to adhere to the surface. This drawing material is not widely used because of the expense of the material. When erased several times, the surface becomes slick, and graphite and ink will not adhere to it. One advantage of using linen sheets is that they are fairly resistant to deterioration when they are stored in a dry place.

A second drawing material is rag content drawing paper. This paper is translucent and has a finish that allows graphite pencils and ink to adhere to the surface. The paper is rather fragile because it is thin and brittle from the finishing process. There are many grades of rag content drawing paper, but even the better quality of material is moderately priced. Rag content drawing paper starts deteriorating rather quickly when stored in a dry place. The sheets become brittle and fall apart after storage for several years.

In recent years, a special plastic drawing film has been introduced on the market. This film is translucent and is a tough plastic-like material that will not tear. It is water-resistant and does not deteriorate for many years. The surface of the film is treated to receive graphite pencil and ink. A special wax-like pencil lead has been marketed that works well on Mylar® film and will not smudge like graphite lead.

8.6 SEPIA DRAWINGS

In Chapter 11 (Section 11.3) the Diazo printing process is discussed, and sensitized print paper is described. Sensitized print paper results from a chemical treatment that is applied to translucent drafting paper. When working drawings are placed over the sensitized paper, a print of the working drawings is made on the drafting paper. This drafting paper is a sepia, which can be drawn on with pencil or ink. Sepias can be printed as working drawings by using the Diazo printing process.

Sepias save time for the drafter, who is not required to trace the floor plan before adding the various systems to be installed in the building. The sepia serves as the floor plan, and the mechanical, plumbing, or electrical system can be drawn directly on the sepia.

8.7 WHO MAKES WORKING DRAWINGS?

The architect and the engineer are responsible for designing a building and preparing working drawings for it. The architect and engineer are registered professionals as required by the state government.

Registered architects and engineers have a seal that is required on all drawings that they are responsible for. This seal must be on the plans before the building can be constructed. (Figure 14.3 is an example of an engineer's seal. The architect's seal is similar.)

After completing their college requirements, architects and engineers must pass a comprehensive series of written examinations and verbal interviews conducted by the state professional examining board. After passing these examinations, they are licensed to practice architecture or engineering in that state. They receive a copy of the bylaws required by that state and have the right to administer the professional seal.

Each state has an examining board; however, the written examinations are prepared and graded by the national examination board. Since the examination is conducted on a national level, a registered architect or engineer can usually obtain a license and seal in another state by reciprocity. The architect and engineer can do design work in other states if they have applied for and received a professional license in the state where the project is to be constructed.

The architect is responsible for building design and for overseeing construction of the building. The engineer is responsible for systems installed in the building such as the plumbing, air-conditioning, heating, and electrical systems.

Designers and drafters who work under the supervision of the architect usually draw the working drawings. The designers, working closely with the architect and the owner, make sketches to show how the building should be constructed. These sketches are given to the drafter who puts the information on the final drawing.

Like the architect, the engineer usually supervises one or more designers and drafters. The engineer and the designers work with the architect to design systems (air-conditioning, heating, plumbing, and electrical) that will serve the building properly and fit in the space set aside for that purpose in the building. The drafter takes the sketches made by the engineer and the designer and makes the finished working drawing.

For related information, the student should refer to Chapter 9, "Categories of Working Drawings" and Chapter 11, "Creating Construction Drawings". Information on freehand sketching and drafting techniques is located in Chapters 24 and 25.

SUMMARY

• Working drawings are the original drawings of plans for a building.

• Copies of the working drawings are called "blueprints" or "construction drawings."

• Working drawings can be created manually or with a computer.

• T-squares, parallel bars, drafting machines, and compasses are used to create working drawings by hand.

• Computer-aided design (CAD) drawings have the advantage of accuracy, consistency, and speed.

• The architect is responsible for the building design and for overseeing construction of the building.

• The engineer is responsible for systems installed in the building.

• Engineers and architects are registered professionals who are assisted by designers and drafters.

• Working drawings are often drawn on linen, rag content paper, or Mylar®.

• Working drawings can be put over sensitized paper and printed as sepia prints.

REVIEW QUESTIONS

1. Define working drawings.

2. What is the difference between working drawings and blueprints?

3. Tell specifically how the following instruments are used to create working drawings:
   A. T-squares
   B. Parallel bars
   C. Drafting machine
   D. Compass
   E. Dividers

4. Why are building plans rarely put on sheets larger than 36" x 42"?

5. What are the components of a computer-aided design (CAD) system?

6. What are the advantages of the preparation of plans using a CAD system?

7. List and discuss briefly each of the steps by which a CAD plan is produced.

8. Discuss briefly the duties of the following in the preparation of working drawings:
   A. Architect
   B. Engineer
   C. Designer
   D. Drafter

9. Define a "sepia." Why are sepias sometimes used for working drawings?

10. Describe some of the kinds of drawing paper used for working drawings, and list advantages and disadvantages of each kind.

9

CATEGORIES OF WORKING DRAWINGS

OBJECTIVES

After studying this chapter, you should be able to:

■ identify and describe the major categories of working drawings

■ discuss the types of plans included under the site work section

■ discuss the types of plans included under the general construction section

■ explain the types of information found in the mechanical section

■ explain the types of information found in the electrical section

■ locate specific items of information in the various subsections of each category of working drawings

9.1 INTRODUCTION

Working drawings are the original drawings prepared by the architect and engineer with the help of their drafters. Chapter 8 discussed the process by which working drawings are created. This chapter is concerned with the categories of working drawings and why working drawings are placed in categories.

9.2 MAJOR CATEGORIES OF WORKING DRAWINGS

Major categories of working drawings are listed below.

1. Site work section

2. General construction section

3. Structural section (can be combined with general construction section)

4. Mechanical section

5. Plumbing section (can be combined with mechanical section)

6. Electrical section

If the construction project is small, several categories can be combined. If the construction job is very large, additional subsections are added to help separate responsibility among the construction contractors.

9.3 SITE WORK SECTION

This section relays information to the contractor who is to do the site work on the construction site. The site plan may consist of a single sheet, several sheets, or many sheets grouped together. The site work section usually consists of the following:

1. **Demolition Site Plan** — shows what exists on the site and what is to be removed (buildings, trees, and unwanted roads, for example).

2. Soil Boring Site Plan — shows the location of test borings to determine types of soils on the site, the location of undesirable rock formations, and the location of undesirable soil.

3. Grading Plan — shows where soil is to be removed or added to make the site suitable for the construction that is planned. This plan also takes into account the drainage of storm water from the site.

4. Site Location Plan — shows the location and dimensions of the buildings on the site.

5. Landscaping Site Plan — shows the location of plantings and lists the names of plants, trees, shrubs, and grasses to be planted.

6. Irrigation Site Plan — shows the irrigation system including piping, location of sprinkler heads, and related equipment for the irrigation system.

7. Utilities Site Plan — shows all service utilities including domestic water, sanitary sewer, electrical, fuel gas, telephone, storm drainage, cable TV, and any other service that is required for the building.

If needed, additional site plans may be added by the architect or engineer to the seven site plans listed above. In many cases, a separate site plan is prepared to show the outside work for a trade area. For example, the site plan for plumbing site work may be included in the plumbing category of plans. Mechanical and electrical categories may also include separate site plans.

9.4 GENERAL CONSTRUCTION SECTION

The general construction category of working drawings includes building construction plans. The general contractor usually does the construction work on small jobs; however, on large jobs, responsibilities may be delegated to other contractors. For example, on a small job, workers hired by the general contractor are usually responsible for the structural or framing portion of the building. On a large job, structural or framing work is done under a separate contract by workers hired by the structural contractor.

A set of general construction drawings is broken down into the following major types of plans:

1. Elevation Plans — show what the building looks like from various viewpoints, including the front elevation, right side elevation, left side elevation, and rear elevation. (See Sheet A-2 of the construction drawings at the end of this textbook for examples.)

In some cases, the building has wings projecting at an angle from the main building. This view of the building will be distorted if it is not parallel to the viewing plane. In such cases, an elevation view is usually drawn to show the wing to scale as seen with a viewing plane parallel to the face of the building. (See Chapter 16 for a detailed explanation of viewing planes.) Elevation views show the relationship of different components such as windows, doors, different textures of building materials, and so forth.

2. Foundation or Basement Plan — shows the location of the foundations for outside walls and columns of the building. (See Sheet 2 of the construction drawings at the end of this textbook for an example of a foundation plan.) If there is a basement, the basement plan shows the size of the basement along with the location of column footings and foundations for the outside walls.

3. Floor Plans — are usually grouped in a sequence beginning with the plan for the first floor and going up a floor at a time to the top of the building. Each floor plan shows the construction features necessary for the completion of work on that floor. Dimensions are given, and symbols for different construction members are incorporated into each floor plan. Elevations, sections, and details are noted on the floor plans. Also noted are instructions telling where these elevations, sections, and details are shown in the set of plans. (See Sheet A-1 of the construction drawings at the end of this textbook for examples of floor plans.)

4. Roof Plans — complete the series of plans for the top portion of the building. The roof plans show all equipment to be located on the roof, roof openings required, roof drains (if roof drains are required), walkways for servicing roof-mounted equipment, and the pitch (or slope) required for the roof. Also indicated are sections and details showing how the roof equipment is to be mounted and how special architectural features are to be installed. The roof plan indicates where these sections and details are located on the plans.

5. Elevations, Sections, and Details — shown on the sheets that follow the roof plan. These drawings, usually done to a large scale, enable the construction worker to understand just how the materials and equipment fit together to meet the requirements of the design. Larger scale drawings allow detailed information and instructions to be included. (See Sheet A-8 of the construction drawings at the end of the textbook for examples of sections.)

6. Equipment Schedules — several equipment schedules are always provided on construction jobs, including the following:

a. Door schedule — shows all doors used on the job. The door schedule shows the door size (dimensions), door type, materials used to construct the door, hardware (hinges and door locks), door finish, door frames, and other information needed for ordering doors.

b. Window schedule — like the door schedule, information required for ordering the windows includes the size, type of glass, type of window, information as to the window frame, and so forth.

c. Room finish schedule — each room in the building is included on this schedule. Such items are listed as the type of ceiling, baseboard, wall treatment (paint or wall covering), and floor finish (hardwood, treated, painted, carpet, etc.). Special notations give instructions on the finishes in each room.

d. Symbol schedule — shows symbols and describes them.

e. Abbreviations schedule — lists the abbreviations used in the general construction area along with an explanation of each abbreviation.

f. Other schedules — include kitchen equipment, hospital equipment, laundry equipment, and special equipment installed in the building under the general construction contract.

Refer to Sheet A-8 of the construction drawings at the end of the textbook for examples of equipment schedules.

9.5 STRUCTURAL SECTION

The structural plans consist of the following:

1. Foundation/Basement Plans — give the size and location for all footings and foundations for the load-bearing wall, outside walls, and columns.

2. Floor Framing Plans — show the structural members that support each floor of the building along with the size and spacing of these structural members.

3. Roof Framing Plans — show the structural framing to support the roof, including size and spacing of structural members. (See Sheet A-8 of the construction drawings at the end of this textbook for examples.)

4. Elevations, Sections, and Details — includes how structural members are to be joined, types of bolts, types of rivets, welds, clips required, and other important information. These drawings are generally drawn to a larger scale to provide information in more detail. (Structural details are shown on Sheet A-8 of the construction drawings at the end of this textbook.)

5. Schedules Plan — schedules on this plan include the following:

a. Column schedule — shows the location, number, type of materials, size, and length of columns.

b. Schedule of miscellaneous steel or structural members.

c. Welding symbols schedule

d. Bolt/nut and rivet schedule

e. Symbol schedule

f. Abbreviation schedule

On small jobs, the architect can show structural work on the architectural plans. Some architects include structural plans at the end of the General Construction section. Most architects group structural plans together in a separate category, particularly on a large job.

9.6 PLUMBING SECTION

The plumbing section is sometimes included as a separate part of the mechanical section. In this chapter, plumbing will be treated as an entirely separate section. The plumbing section usually includes the foundation or basement, floor plan, roof plan, elevations, sections and details, and equipment schedules.

Plumbing work outside the building is shown on the site plan. This site plan is usually bound with the other site plans as described in section 9.3 above.

Sometimes, the plumbing site plan is bound as the first sheet of the plumbing section.

In addition to the plumbing work site plan, the plumbing section usually includes the following:

1. The Foundation Plan — shows piping that is installed under the first floor. The piping is located in the crawl space (if there is one), or under the floor slab of a concrete floor on grade. Piping consists of hot and cold domestic water, sanitary sewer or drainage piping, and special services piping such as fuel gases, medical gases, or process piping.

2. Basement Plan — piping serving the basement (underfloor piping) is shown on a basement floor plan if there is a basement; however, notes on the plan specify whether the piping is located under the basement floor or at the basement ceiling level.

3. Floor Plans — each floor plan shows plumbing fixtures and equipment served by the plumbing system. Each piece of equipment and/or plumbing fixture has a symbol referring the print reader to the schedule section to see exactly what equipment/fixture is specified.

Piping serving the plumbing equipment/fixtures is usually shown on the floor plan where the piping is to be installed. In other words, if the piping is located above the second floor ceiling and serves equipment/fixtures on the third floor, the piping is drawn on the second floor, showing all runouts and riser piping that turn up to the third floor. The plan is noted to say that piping shown on this plan is to be located above the second floor ceiling. In some cases, the engineer/drafter shows the piping for the equipment/fixtures on the same floor plan on which the equipment/fixtures are shown. To prevent confusion, the engineer/drafter adds notes that explain where the piping is to be installed.

4. Roof Plan — all plumbing work required on the roof or through the roof is shown on the roofing plan. This includes plumbing vents through the roof, roof drains (if required), gutters (if required), and special plumbing equipment to be mounted on the roof.

5. Elevations, Sections, and Details Plans — these drawings, usually drawn to a larger scale, show details and give information necessary for proper installation of the plumbing system, equipment, and fixtures. An isometrically drawn piping diagram is not uncommon for the detail sheet. (See Chapter 18, "Isometric Drawings," for a detailed explanation of isometric diagrams.)

6. Equipment and Fixture Schedules — schedule sheets usually contain the schedules for plumbing fixtures, plumbing equipment, piping, piping connections, plumbing symbols, and plumbing abbreviations.

Sheet P-1 of the construction drawings at the end of this textbook shows the plumbing floor plan, details, and the equipment and fixture schedule for a small bank and trust company.

9.7 THE MECHANICAL PLAN SECTION

The mechanical section, like the plumbing section, may include a site plan at the beginning showing mechanical work to be done outside the building.

In addition to this site plan, the mechanical section includes the following:

1. Foundation and/or Basement Plans — show mechanical work to be done in the crawl space, under the slab on grade, or in the basement.

2. Floor Plans — one plan per floor shows mechanical work required for each floor. Ductwork is shown on the floor where it is to be installed. For example, ductwork located above the second floor ceiling and serving the second floor is shown on the second floor plan. If ductwork is located above the ceiling of the second floor to serve a space on the third floor, the ductwork is drawn on the second floor and is noted as being installed on the second floor. Piping for the heating and cooling system is drawn and noted in the same manner.

3. Elevations, Sections, and Details — located on sheets that follow the floor plan sheets. These large scale drawings show how units, ductwork, and piping are installed. The drawings also show how the units are to be connected with piping, ductwork, water, steam, and so forth.

4. Schedules — mechanical schedule sheets such as those listed below relay information to the mechanical worker:

    a. Unit or equipment schedules

    b. Register, grilles, and diffuser schedules

    c. Duct insulation schedule

    d. Piping schedule

Sheet M-1 of the Construction Drawings at the end of the textbook shows mechanical plans for a small bank and trust company, including ductwork, piping installation, and sections and details.

9.8   ELECTRICAL SECTION

Electrical work shown on the site plan can be the first sheet of the electrical section. The electrical section also includes the following:

1. Foundation/Basement Plan — shows electrical equipment (panel boards, switches, electrical outlets, etc.) required for the basement, located in the crawl space, or located under a slab on grade.

2. Floor Plans — show the electrical equipment required for each floor, including receptacles, lighting fixtures, and necessary electrical connections to equipment furnished by other contractors.

3. Roof Plan — electrical wiring, equipment, and electrical connections to equipment on the roof are shown on the electrical roof plan.

4. Elevations, Sections, and Details — these are detailed drawings showing how the electrical equipment is installed. Special instructions and information are relayed to the electrical worker through these drawings.

5. Schedule Plan — contains schedules for the electrical devices, including the following:

    a. Lighting fixture schedule

    b. Panel board schedule

    c. Conduit and raceway schedule

Sheet E-1 of the construction drawings at the end of this textbook shows the electrical floor plan, symbol schedule, lighting fixture schedule, and a panel board schedule for a small bank and trust company.

SUMMARY

• Working drawings are original drawings prepared by the architect and engineer to give information to the contractor on a construction job.

• Working drawings are usually classified in the following categories: site work, general construction, structural, mechanical, plumbing, and electrical.

• Site work plans give details on the preparation of the site for construction.

• General construction plans include the foundation plans; floor plans; roof plans; elevations, sections, and details; and equipment schedules.

• Structural plans give information to contractors on the foundation, framing, and other structural information needed for erecting the building.

• Plumbing plans are sometimes shown as part of the mechanical section.

• The mechanical section contains information about piping, ductwork, and systems to be installed in the building.

• The electrical section contains details on panel boards, electrical outlets, and electrical fixtures to be installed in the building.

REVIEW QUESTIONS

1. Define working drawings.

2. Discuss briefly each of the following types of site work plans:
   A. Demolition site plans
   B. Soil boring site plans
   C. Grading plans
   D. Site location plans
   E. Landscaping site plans
   F. Irrigation site plans
   G. Utilities site plans

3. What types of elevation plans are usually included in the general construction section?

4. What is the sequence for floor plans, and what information is found on them?

5. What kinds of information are shown on the roof plans?

6. Why are elevations, sections, and details usually drawn to a larger scale than other parts of a set of plans?

7. What is the purpose of the Door Schedule?

8. What information is found in the structural section?

9. What types of piping may be shown on the foundation plan of the plumbing section?

10. How does the architect/engineer indicate on the floor plans that piping above the second floor ceiling serves fixtures on the third floor?

11. Where should a person look on the mechanical plans for specific information on how to install units, ductwork, and piping?

12. What is the purpose of the schedule plan in the electrical section?

13. Describe where each of the following would be found:
    A. Which trees are to be removed
    B. The location of columns of a building
    C. The slope of the roof
    D. The kind of wall covering
    E. The size and style of doors
    F. How the roof is to be framed
    G. If piping is to be located under a floor slab
    H. Whether gutters are required
    I. Where ductwork serving a particular floor is to be installed
    J. What shrubs and trees are to be planted

10

SECTIONS, ELEVATIONS, AND DETAILS

OBJECTIVES

After studying this chapter, you should be able to:

- ■ recognize and define sections, elevations, and details

- ■ discuss reasons why sections, elevations, and details are shown on plans

- ■ use sections, elevations, and details to find specific information about building construction

10.1 INTRODUCTION

Sections, elevations, and details are an integral part of any set of construction drawings. They are used by architects and engineers to relay information that cannot be shown on the floor plans. This information helps building construction workers understand how the building is put together and how to construct it as designed. HVAC technicians use the information for calculating heating and cooling loads and for designing and installing systems that will fit into the allotted spaces.

Sometimes the architect and engineer put sections, elevations, and details on separate sheets; however, on small jobs these drawings are put on unused spaces of the construction drawings for the various trades. Architectural plans contain details pertaining to construction, mechanical plans have details relating to the HVAC system, and so on for each trade involved in the project.

10.2 SECTIONS

Section drawings are drawn to scale, showing internal portions of objects. Section drawings are developed through the use of a cutting plane.

The cutting plane is an imaginary single plane similar to a pane of glass. This cutting plane is used to cut (visually) through a wall, mechanical unit, or other object. In Figure 10.1, the cutting plane is inserted into an apple. In Figure 10.2, one-half of the apple has been removed, and the inside of the

Figure 10.1 Cutting plane

apple can be seen through the cutting plane. This view of the apple is the section view.

Sections show how the components inside a drawing relate to each other. Construction sections enable the builder to see how different members of a building are joined and their relationship to each other. Figure 10.3 shows a section of a cornice. All of the various construction members are shown

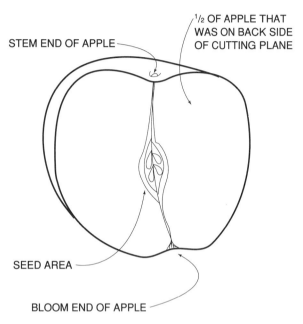

Figure 10.2 Section view of an apple

and named. Sometimes dimensions are included on section drawings.

Figure 10.4 is a section view of an air-conditioning unit. The duct connection, supply fan, cooling coil, filter section, and return air duct connection can be seen. The internal condensate pan is shown along with flexible duct connections on the return and supply sides of the unit.

Sections are drawn for any portion of work in each trade to clarify how a unit functions or how components are joined. A wall section with footings, insulation, flooring, and other components is shown in Figure 10.5. Since wall sections are usually drawn to scale, dimensions have been omitted for clarity. Figure 10.6 is a drawing of a partial wall section with stucco on the outside of the wall and paneling on the inside.

Arrows or special markings representing the cutting plane identify the point at which sections are taken. Figure 10.7 is a plan view of typical locations where sections are cut. Two types of symbols indicate the cutting plane. The first is a heavy line drawn into the object with a ninety degree line, an arrowhead, and the letter "A" shown at the point of the arrowhead. This symbol indicates the location of the cutting plane and the direction one should look to see the section. Using this symbol, Figure 10.7 identifies the section as Section A-A. If there are

Figure 10.3 Section view of a cornice for a brick veneer wall

FILTER SECTION WITH 2" FILTERS

FLEXIBLE DUCT CONNECTION

SUPPLY AIR FAN

SUPPLY AIR DUCT

RETURN
AIR DUCT

CONDENSATE DRAIN PAN
UNDER COIL

COOLING COIL

FLEXIBLE DUCT CONNECTION

Figure 10.4   Section of an air-conditioning unit

SHINGLES OVER ROOFING FELT

PLYWOOD SHEATHING

INSULATION

ROOF BEAM OR RAFTER

SHEATHING

TOP PLATE

SIDING

STUD

DRY WALL

BASE BOARD

MOLDING

INSULATION

HARDWOOD FLOORING

SOLE

BLOCKING

PLYWOOD SUBFLOOR

HEADER

FLOOR BEAM OR JOIST

TERMITE SHIELD

CRAWL SPACE

FINISHED GRADE

SILL

UNDISTURBED EARTH
UNDER FOOTING

CONCRETE FOOTING

Figure 10.5   Wall section

WOOD PANELING
ON ⁷/₈" GROUNDING

¹/₂" PLASTER ON LATH

2 x 4 STUDS

⁷/₈" SHEATHING

BUILDING PAPER

³/₈" WOOD FURRING

1" STUCCO ON
WIRE LATH

³/₈" SELF-FURRING
WIRE LATH

Figure 10.6 Partial wall section

other sections, they will be identified as Section B-B, C-C, and so on.

A second type of symbol indicating the cutting plane is also shown on Figure 10.7. A heavy line is drawn outside the object to a circle which has a number above the center line and a combination letter and number under the line. An arrowhead formed on the circle indicates which way one should look to see the section. The number above the center line indicates the number of the section. The combination letter and number under the line gives the sheet number where the section is drawn.

Using the combination symbol, Figure 10.7 identifies the section as Section 3 shown on Sheet M-1. Sheet A-2 of plans at the end of this textbook demonstrates sections that are cut through building elevations. Sheet A-5 of the same set of drawings contains longitudinal sections and transverse sections through the building. Drawings of this type allow the construction worker to visualize what the building will look like and provide information that helps the HVAC designer locate equipment.

PLAN VIEW

SECTION A-A
OR
SECTION 3 SHOWING SHEET M-1

Figure 10.7 Two ways to show location of section

10.3 ELEVATIONS

Elevations are scaled drawings showing what the building should look like when construction is completed. Elevations are drawn on a single plane, showing no depth. They supply information that cannot be given on the plan view of an object. Elevation drawings tell locations of windows and doors, types of wall finishes to be used, exterior trim, location of exterior cooling and heating equipment, and other vital information. Typically, building elevations are shown for each face of the building (front side, left side, right side, and rear). Figure 10.8 shows a simple building elevation. Several examples of exterior building elevations are shown on Sheet A-2 at the end of this book.

Larger elevations can be drawn starting with the plan view of an object. The finished plan view is taped on the sheet where the elevation is to be drawn. Light extension lines are drawn between the plan view and the elevation location. These extension lines transfer the necessary horizontal distances from the plan view to the elevation view. The drafter uses the architect's scale to draw the height dimensions on the elevation. This method saves time and allows the drafter to draw the ele-

vation drawings accurately. (See Figure 10.12 for a plan view of a condensing unit with extension lines drawn between the plan view and the elevation view.)

Interior wall elevations are sometimes drawn to show the wall from floor to ceiling and from adjoining wall to adjoining wall. The drafter labels these drawings as wall elevations. (See Sheet A-7 of the plans at the end of this textbook for examples of wall elevations. These drawings are different from the section drawings shown on Sheet A-5 because the section covers the entire building and a wall elevation covers only one wall inside the building.)

10.4 DETAILS

At times, the architect and engineer need a drawing that gives a three-dimensional appearance. These drawings are labeled "Detail" on the plans under the title of the detail. Details are not usually drawn to scale. For clarity, the drafter notes "Not to Scale" or gives the scale used. Some examples of such drawings are shown in Figures 10.9 through 10.13, pages 56–58. Each of these details gives a realistic view of the object along with specific necessary information.

BRICK CHIMNEY

GERMAN SIDING

COMPOSITION SHINGLES

5 X 3 DOUBLE HUNG WINDOWS
(THERMO-PANE)

FOUNDATION

THERMO-PANE
6 X 5 PICTURE WINDOW

2 X 5 DOUBLE HUNG WINDOW
(THERMO-PANE)

Figure 10.8 Front elevation of building

(*NOTE:* See also *Chapter 17, "Oblique Drawings,"* and *Chapter 18, "Isometric Drawings".*)

SUMMARY

• Details, sections, and elevations are included on plans to relay information from the architect to persons who will be studying the construction drawings.

• Details, sections, and elevations may be on a separate page or on unused spaces on the construction drawings.

• A detail shows how materials and components are joined, and it is not usually drawn *to scale*. Details are often drawn using the isometric or oblique drafting method.

• A section drawing shows the internal portion of an object and is usually drawn *to scale* using a cutting plane. Sections illustrate how different components of a building are joined and how they relate to each other.

• An elevation shows in detail what an object actually looks like and is usually drawn *to scale*. Elevations usually show the different views of a building and give detailed information about its parts, including locations and (sometimes) dimensions.

REVIEW QUESTIONS

1. Define a detail, and describe its purpose.

2. Define a section, and describe its purpose.

3. Define a cutting plane, and show an example of a drawing using a cutting plane.

4. Define an elevation, and describe its purpose.

5. How does a drafter indicate whether a detail has been drawn to scale or not to scale?

6. List examples of information usually found on elevation drawings.

ADDITIONAL STUDENT EXERCISE: EXTRA CHAPTER 10 REVIEW

NOTE: Refer to Figures 10.9 through 10.13 to answer the following review questions.

1. Figure 10.9 is a detail of a hot water unit heater. How many globe valves are shown?

2. Describe the water flow through the unit heater in Figure 10.9.

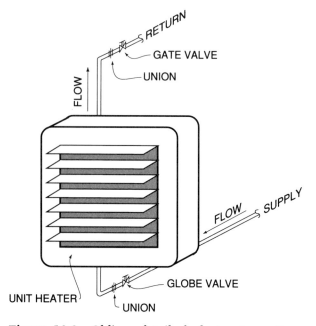

Figure 10.9 Oblique detail of a hot water unit heater

3. In Figure 10.10, is the filter located inside the furnace or outside the furnace?

4. How high is the concrete housekeeping pad in Figure 10.10?

NOTE: NOT TO SCALE

Figure 10.10   Detail of connections to a gas furnace in an A/C unit

5. Is any provision made for venting air from the coil in Figure 10.11?

6. Name the types of valves shown in Figure 10.11.

Figure 10.11   Detail of hot water heating coil with three-way control valve

7. In Figure 10.12, what information on the elevation view is not shown in the plan view?

8. How high is the mounting pad in Figure 10.13?

9. List the information shown in Figure 10.13 that is not shown in Figure 10.12.

10. Looking at Figures 10.12 and 10.13, which detail do you prefer? Give reasons for your choice.

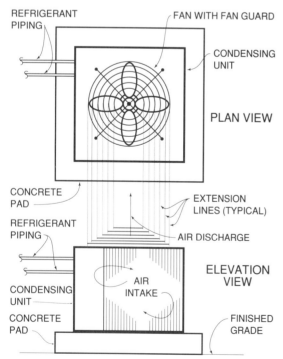

Figure 10.12   Plan view and elevation view of condensing unit

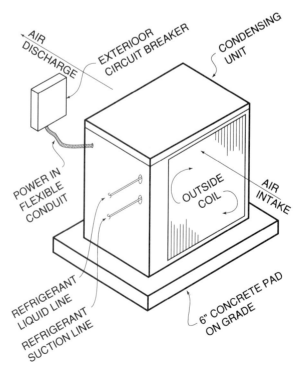

Figure 10.13   Detail of outside condensing unit

11

CREATING CONSTRUCTION DRAWINGS

OBJECTIVES

After studying this chapter, you should be able to:

- explain how the sensitized iron salts process is used to create blueprints
- list disadvantages of using the sensitized iron salts process
- explain how ultraviolet light is used in the Diazo process
- outline the major steps in making construction drawings with the electrostatic printing process
- list advantages and disadvantages of the electrostatic printing process
- describe the way CAD machines can be used to make construction drawings
- discuss the use of printers in the creation of construction drawings

11.1 INTRODUCTION

Construction drawings are the plans that are used by the construction crew on the job site. Construction drawings, commonly called "blueprints," are reproductions of working drawings. Chapter 8 discussed how working drawings are created using conventional drafting and computer-aided drafting (CAD). Working drawings are then reproduced by various methods to become construction drawings. The discussion in this chapter will deal with the reproducing process.

11.2 HOW CONSTRUCTION DRAWINGS GOT THE NAME "BLUEPRINT"

As discussed in Chapter 8, it is not practical to use working drawings as construction drawings. Working drawings are made on translucent drawing film that is usually very fragile. These drawings could not withstand use on the job site.

Years ago, working drawings were drawn on a heavy paper or linen. The architect and engineer drew the original designs or drawings and then hired drafters to copy them on heavy paper or linen. (See Chapter 8 for a description of drawing paper or film.) The process was costly, and only a few sets of drawings were produced to construct a building.

The first "blueprint" process for reproducing drawings was discovered around 1840. This process enabled the architect and engineer to make copies of the working drawings produced in their offices. The process consisted of sensitizing a heavy paper with iron salts. This paper underwent a chemical change when exposed to light.

First, the sensitized paper was placed on a pane of glass. Then the architect's or engineer's working drawing (drawn on translucent paper) was placed over the sensitized paper. A second pane of glass was placed over the working drawing sheet. These pieces of glass were placed in a frame to hold the glass and paper firmly in place (Figure 11.1). The frame was then exposed to sunlight for a specified

Figure 11.1 Original method for exposing prints to light — Blueprint

There were other disadvantages to the blueprint process. Since the sensitized paper had to be washed with water, washed in a potassium dichromate, and then run through a final water rinse, the print was stretched out of shape when it dried. After drying, the paper was wrinkled and very brittle. Because of the wet baths and then drying, the drawings were stretched to a point that a scale used on the blueprint gave only an estimate of the true measurement. For accurate measurement, one had to rely on the dimensions given on the plan. Very few blueprints are made today using the sensitized iron salts process; however, if blueprints are required, they can be produced by machine. Figure 11.2 shows a block diagram of the various printing processes for producing construction drawings from working drawings. A brief discussion of each of the processes follows.

11.3 THE DIAZO PROCESS FOR MAKING CONSTRUCTION DRAWINGS

The most common method of reproducing working drawings for construction drawings is the *Diazo process*. The Diazo process makes prints with white

time. The sunlight caused a chemical reaction on the sensitized paper. Where there were lines on the working drawing, the sunlight was prevented from reacting with the iron salts.

After exposure to sunlight, the sensitized paper was washed with water. The area of the paper that was not protected from sunlight by lines on the working drawing turned blue. The area protected from sunlight by lines on the working drawings turned white.

The sensitized paper was then run through a second bath containing potassium dichromate to fix the color. A final rinse with water finished the process. Then the sensitized paper was spread out to dry. The blue color of these reproduction drawings gave them the name "blueprints."

This method of reproducing plans was a trying and time-consuming task. If the paper was exposed too long to sunlight, the reproduction bleached out so much that the print was not readable. If not enough exposure to sunlight was made, the print was too blue and unreadable. Even with these drawbacks, the reproducing process was far faster than tracing the working drawings over and over.

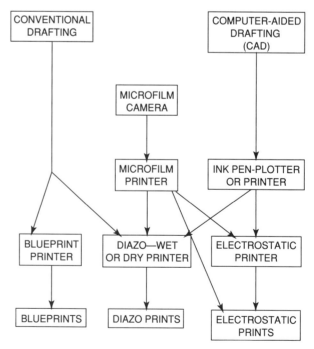

Figure 11.2 Paths for making prints today

backgrounds and black lines or white backgrounds with blue lines. As a rule, architects prefer the white background with black lines, and engineers prefer white backgrounds with blue lines.

The Diazo process uses a heavy duty, sensitized paper for printing. This paper is coated with a chemical that reacts to ultraviolet light. A specially designed machine is used to reproduce prints at a fairly fast rate (Figure 11.3).

The Diazo print machine consists of one section that exposes the sensitized paper to ultraviolet light and a second section that develops the print. The exposure section is driven by a variable speed motor that limits the time the paper is exposed to the light. Larger machines have a series of conveyor belts that rotate around a clear cylinder inside the machine. Several fluorescent tube lights that emit ultraviolet light are located inside the clear cylinder. Figure 11.4 is a diagram of the path that a sheet of copy paper follows in a typical Diazo print machine.

Sensitized paper is placed on the work surface of the machine with the chemical (yellow) side up. The working drawing that has been drawn on translucent paper is placed on top of the sensitized paper. The two papers are then pushed into the machine and around the clear cylinder, exposing the sensitized paper to the ultraviolet light. The light passes through the working drawing, and areas not protected by lines on the working drawing are exposed to the ultraviolet light. Areas that are protected by lines on the working drawing are not exposed to the ultraviolet light.

Figure 11.3 Printing machine *(Courtesy of BluRay, Incorporated)*

When the exposure process is complete for the two pieces of paper, the working drawing is discharged from the machine. The machine operator then inserts the sensitized paper into the developing section of the machine where it is developed into a finished construction plan.

The Diazo printing machine is manufactured in many sizes and styles. Figure 11.5 is a picture of a table model of a typical Diazo machine often used in architectural and engineering offices. Much larger machines are also available that develop prints at a rate of 16 feet per second or higher. These machines are used in architectural and engineering offices and in commercial printing shops where a large volume of prints is required.

Diazo prints have been proven to withstand heavy use on the job. The process produces a good quality print with crisp lines and a good contrast between the lines and background.

There are, however, two disadvantages to Diazo prints. If these prints are exposed to direct sunlight for a period of time, they will bleach out to a white paper. If the prints are exposed to water, the blue or black lines will run on the background.

11.4 THE ELECTROSTATIC PRINTING PROCESS

The electrostatic printing process, which has only recently become available, is very popular. The printing process is the same as the process used in the office copier. Since the Xerox Company was one of the first to put an office copier machine on the market, the name Xerox is commonly (although erroneously) used to refer to all office copiers.

The electrostatic printing machine can print full-size copies (24″ x 36″) of the original working drawings prepared by the architect and engineer on microfilm cards containing the image of the original working drawings (Figure 11.6).

In the electrostatic printing process, the working drawing is exposed to light, and the image of the working drawing is projected through a lens to a negatively charged drum inside the machine. The drum is discharged by the projected light, and areas of the drum protected by lines on the working drawing remain negatively charged. Then the drum is rotated past a roller where black toner particles are

Figure 11.4 Path of Diazo printing machine *(Diagram courtesy of BluRay, Incorporated)*

attracted to the negatively charged areas on the large drum surface. As the drum continues to rotate, it comes into contact with the print paper, which is positively charged. The toner particles are attracted to the positively charged print paper. The print paper, with the toner particles attached, is then passed through two rollers where heat and pressure fuse the toner to the print paper.

Prints produced by this method are easily read. The background is white and the lines are black. The quality is the same as that obtained from a good office copier. The print paper is usually fairly heavy, of good quality, and it holds up well on the construction site. When this process is used, the print can also be enlarged or reduced with the electrostatic printing machine. The features are not available with the blueprint or the Diazo processes.

Construction prints produced by the Diazo process and the electrostatic process, as noted for the blueprint process, should not be scaled to determine distances. In the Diazo process, the print paper and the working drawing are rotated over a large cylinder to establish the image on the print paper. Rotation over the cylinder distorts the actual image on the print paper so that the image is no longer the same

as the one that was drawn on the working drawing. With the electrostatic printing process, the image is distorted as the lines and image are projected on the negatively charged drum.

Electrostatic prints are gaining in popularity in the industry. In some cases, this type of printing costs more than the Diazo process, and cost is sometimes a factor. The quality of the print is usually very high with sharp lines and a good contrast between the lines and the background. The paper is about the same quality as that used in the Diazo process, and it withstands heavy use on the job. When exposed to sunlight, the electrostatic print does not fade as quickly as the Diazo process print. When wet, the paper wrinkles, but the lines do not run like the Diazo process print.

All in all, construction drawings produced by both the Diazo process and the electrostatic process are

Figure 11.5 Table model of Diazo printing machine *(Courtesy of BluRay, Incorporated)*

Figure 11.6 Electrostatic printing machine *(Courtesy of Hewlett-Packard Company)*

of good quality. The architect and engineer have choices as to the process used for making construction drawings to relay information to the craftsmen on the job.

11.5 OTHER METHODS OF PRINTING CONSTRUCTION DRAWINGS

Another method of printing construction drawings is gaining in popularity. This method is the printing process used with the computer-aided design (CAD) machine. If a one-page drawing is needed on the job to clarify a construction detail, the architect or engineer sometimes does the drawing on the CAD computer and prints the drawing with the pen plotter that is driven by the CAD computer (see Figure 8.5, page 42). The drawing is printed on a heavier grade paper than is usually used for working drawings. The paper has a white background with lines that have been drawn in black ink. One disadvantage of a plan created in this manner is that it is not durable enough for heavy use on the job site.

The CAD computer also prints plans with a dot matrix printer that operates similar to a typewriter. Rows of pins strike a carbon ribbon, transferring carbon to the paper and creating lines, letters, and symbols on the drawing. These plans have a white background with black lines, and the quality is similar to that of a typewritten letter. These prints are more durable than those created with a pen plotter. Like pen plotter prints, dot matrix prints are not used to a great extent as construction prints because they will not withstand heavy use on the job site.

Laser printers similar to the office copier are also available for use with the CAD computer. Electronically transmitted information is transmitted from the computer to the printer paper. This transmitted information attracts toner particles to the print paper to form the drawing image on the paper. Toner particles are then fused to the paper with heat and pressure. Prints made by the laser printer have black lines on a white background. These prints are more permanent than the pen plotter prints; however, they are used on a very limited scale as construction prints.

Prints made by CAD and the pen plotter, dot matrix printer, and laser printer are usually restricted to the creation of working drawings that can be reproduced by the Diazo printing method or the electrostatic printing method described above. The pen plotter, dot matrix printer, and laser printers were designed and are used for drawing on translucent drawing paper or translucent drawing film. (See Chapter 8, "How Working Drawings Are Created.")

11.6 MICROFILM AND COMPUTER DISK STORAGE OF DRAWINGS

To save space and for ease in filing record prints, architects and engineers often use a microfilm camera and microfilm printer. The microfilm printer prints the drawing on a standard size drawing sheet or drawing film which can then be printed on a heavier paper using the Diazo or electrostatic process.

Using CAD also offers an opportunity to save drawings using a minimum amount of space. A CAD machine can copy drawings onto a computer disk, which can be stored for future reference or to make future copies of the working drawings. The amount of storage space needed for this system is much less than is needed for the old way of storing full-size working drawings.

SUMMARY

• Construction drawings are reproductions of working drawings.

• Construction drawings must be durable enough to withstand heavy use on the job site.

• "Blueprints" are made by treating a heavy paper with iron salts, exposing it to sunlight, and subjecting it to the developing processes.

• The blueprinting process is more time-consuming and less satisfactory than processes that have been developed more recently.

• The most common process used today to reproduce construction drawings is the Diazo process, which exposes sensitized paper to ultraviolet light and then uses a developing method to complete the process.

• The electrostatic printing machine uses a process like that of the office copier.

• Prints made with the electrostatic printing machine can be enlarged or reduced in size.

• Drawings can be reproduced with a pen plotter, a dot matrix printer, or a laser printer used in conjunction with a CAD machine.

• Microfilm and disk copies save space and are a convenient way to store working drawings in the office.

REVIEW QUESTIONS

1. Define construction drawings.

2. Why are the original working drawings not used on the job site?

3. Discuss the major steps in the blueprinting process.

4. What are three disadvantages of the blueprinting process?

5. What is the most common method used today for reproducing drawings?

6. Outline the steps followed in reproducing prints by the Diazo process.

7. Describe the electrostatic printing process.

8. What are three advantages of the electrostatic printing process?

9. Make a chart on which you compare the advantages and disadvantages of each of the processes for reproducing construction drawings that have been discussed in this chapter.

10. Describe how the CAD machine may be used in conjunction with other machines to reproduce prints.

11. What is the advantage of saving working drawings on microfilm in the architectural or engineering office?

12

WHO USES CONSTRUCTION DRAWINGS?

OBJECTIVES

After studying this chapter, you should be able to:

- discuss the process and order by which various persons and agencies review construction drawings
- tell why construction review agencies have the authority to deny permission or stop construction of a building
- define "invited bidding" and "open bidding" of contracts
- define and tell the importance of a "bid bond"
- tell how equipment suppliers determine their bids
- discuss reasons why record drawings are important

12.1 INTRODUCTION

Preceding chapters have discussed how working drawings are made, categories of working drawings, and how construction drawings are made from working drawings. This chapter will focus on the persons who use construction drawings.

When working drawings have been completed by the architect or engineer, several sets are printed and distributed as construction drawings to the owner and to various companies and agencies for review and approval. In this chapter, the discussion will progress in the same manner as the usual approval process, ending with record drawings (a form of construction drawings) being turned over to the owner at the end of construction.

12.2 PLAN REVIEW AND APPROVAL BY THE OWNER

When a set of construction drawings has been completed, it is presented to the owner for a thorough review. During this review, the owner presents questions to the architect and engineer. Any changes to be made are directed to the architect and engineer. When changes have been made on the working drawings (see Chapter 8), the corrected construction drawings are printed (see Chapter 11). Afterward, they are again presented to the owners for review and approval.

12.3 REVIEW AND APPROVAL BY BUILDING CONSTRUCTION AGENCIES

After being reviewed by the owner, the construction drawings are submitted to local and state building construction review agencies for review and approval. These agencies have been granted the responsibility and authority by the local and state government to require changes in the plans and, if necessary, to deny permission to build. The building construction review agencies review the drawings to verify that the design, materials, methods of construction, location of building, and all aspects of the drawings and specifications comply with the requirements and intent of the building codes of the area (see Chapter 13). This review may take several weeks to complete; however, no construction can start before the review has been made and plans have been approved.

The architect or engineer makes any corrections or changes necessary to secure the approval of the reviewing agencies. Afterward, a building permit can be issued and construction can start. A set of corrected drawings again goes to the owner for final review. When final approval is received from the owner, the architect and/or engineer is ready to print the construction drawings and proceed with the next stage of the building process.

12.4 CONTRACTORS FOR BIDDING (PRICING)

In most cases, construction plans are issued to several construction companies for competitive bidding. The list of bidding companies is agreed on by the owner and the architect/engineer. This process is called *invited bidding*. Companies on the bid list are the only ones allowed to submit a construction bid. On construction projects that are financed with public money (taxes), the bid list is usually open. An *open bid list* means that any company large enough to do the construction is allowed to bid the job.

A *bid bond* is usually required for construction projects. This bid bond is an insurance policy or available security offered by the bidding company to cover liability for the bid if the bidding company should not be able to complete the work. In that case the insurance company or security company would hire another contractor to finish the construction project.

Plans are issued to contractors in all areas of construction (building construction contractors, plumbing contractors, mechanical contractors, and electrical contractors, for example).

12.5 EQUIPMENT SUPPLIERS

Equipment pricing is important for determining the amount of a bid in any area of construction. Equipment suppliers need access to construction plans to be able to price the specified equipment. In most cases, these suppliers contact the bidding contractors and use the contractors' plans for "taking off" the equipment for pricing. "Taking off" equipment means listing measurements, number, and kinds of equipment needed for the job. (It is not uncommon for each bidding contractor to request and receive several sets of plans for workers who will do the take-off and pricing.)

12.6 CONSTRUCTION PLANS FOR CONSTRUCTION CONTRACTORS

When the bids have been taken and the owner has issued a contract to successful bidders, the architect/engineer issues the specified number of sets of plans to the contractor. The specified number of plans for construction is usually listed in the specifications (see Chapter 13). Contractors issue the plans to the foremen who will do the work on the site.

During construction, the job is inspected periodically by employees of the Building Inspection Department to see that work is being done properly. These periodic inspections are very important because the Building Inspection Department has the authority to stop construction until incorrect work is corrected.

12.7 RECORD DRAWINGS MADE FROM CONSTRUCTION DRAWINGS

During construction of the building, certain changes are almost always necessary to accommodate the installation of various materials and equipment. It is almost impossible to coordinate every piece of pipe, every section of ductwork, and every piece of conduit so that there are no conflicts. The changes necessary to eliminate these conflicts are made on the job, and the contractor is usually required to keep a set of construction drawings marked up to reflect these changes.

At the end of the job, the marked-up construction drawings are turned over to the architect/engineer for correction to show changes made on the job. After these corrections have been made, the architect/engineer runs several sets of plans of the finished project for the owner's use. Since these plans give the owner a record of how the building was constructed and the locations of various components of the finished construction, they are called *record drawings*. The record drawings will be used for reference in case of future renovation of the building, for making repairs, and for identifying piping and valve locations, gas service piping, and other mechanical and plumbing services.

The number of corrected construction drawings turned over to the owner should comply with the agreement made between the owner and the architect/engineer at the beginning of the design.

SUMMARY

• Copies of working drawings issued to the owner, contractors, and review agencies are called construction drawings.

• The owner has the opportunity to have changes made in the construction drawings before and after the review agencies approve them.

• No construction can begin until plans have been approved and a building permit has been issued by agencies established for that purpose.

• Inspectors for the building construction agencies can stop construction if work is not done properly.

• "Invited bidding" means that only those companies on the "bid list" can bid on a job.

• "Open bidding" means that any contractor large enough to do the job may submit a bid.

• Construction drawings are needed by equipment suppliers who are submitting bids on a job.

• All contractors on a job are issued copies of the construction drawings.

• Any changes made during construction are recorded on the construction drawings, which are issued to the owner at the completion of the job as "record drawings."

REVIEW QUESTIONS

1. Discuss briefly the persons, agencies, and contractors who may ask for changes in the construction drawings before the building is constructed.

2. Why does the architect or engineer go over the construction drawings with the owner at several points in the review process?

3. Why is it important that the building construction review agencies react favorably to the construction drawings?

4. What authority does the inspector hired by the construction review agency have?

5. Define:
   a. open bidding
   b. invited bidding
   c. bid bond
   d. "taking off" equipment for pricing

6. How does the architect/engineer usually determine how many sets of construction drawings are to be issued to various contractors?

7. What are record drawings?

8. Why is it important for the owner to have a correct set of record drawings?

13

SPECIFICATIONS

OBJECTIVES

After studying this chapter, you should be able to:

■ explain why specifications are an important part of a set of plans

■ tell how specifications are usually organized

■ give examples of information found under Division 1 — General Requirements

■ tell what types of work are described under Division 2 — Site Work

■ find specific information listed under each of the other divisions on a set of specifications

13.1 INTRODUCTION

Specifications contain written information that is not shown on the plans but is necessary for the completion of the job. Specifications contain information on (1) procedures for starting the job, (2) meetings planned to schedule phases of construction, (3) how payments by the owner are to be made, (4) where equipment is to be stored until needed, (5) insurance that must be carried by the contractors, and (6) other details. On very small jobs where specifications consist of some ten or twelve pages, architects may print them on a plan sheet rather than binding them in a separate book that is included with a set of plans. (See Sheet M-2 of plans at the end of this textbook for an example of specifications printed on a plan sheet.) In most cases, the specifications are written, printed, and bound into a book. Specifications are often referred to as "specs" in the office and on the job.

The length of the specifications varies with the size and type of construction. If the job is small and simple, and if the construction is standard procedure, the specifications can be short. If the job is large and has many special features and necessary procedures, the description of procedures may be lengthy. Specifications written for federal government jobs are known for being long and detailed. The government dictates what is to be included in the specifications, and the specifications become elaborate because each item is described in great detail.

In this chapter, the major divisions of a set of standard specifications will be listed, along with a brief discussion of each division. There are sixteen major divisions of specifications; however, some architects and engineers consider plumbing as a separate division rather than grouping it with mechanical (Division 15). In that case, plumbing becomes Division 16 and electrical becomes Division 17.

13.2 DIVISION 1 — GENERAL REQUIREMENTS

This division includes general administrative and technical provisions that may not be listed elsewhere and which may apply to more than one division. General requirements include contractual/legal requirements, a summary of work to be done, an explanation of work to be performed later, a description of meetings to be held (construction coordination meetings, scheduling meetings, and progress meetings, for instance), quality control, and submittal. "Submittal" refers to catalog descriptions or "shop drawings" of equipment, which are submitted by the contractor for approval by the architect and/or engineer.

Division 1 also includes a description of temporary facilities, preparation of the job site for construction, and details how the job is to be closed out when construction is completed. Listed in the General Requirements are such details as how many sets of construction plans are to be furnished to the construction contractor for the project.

Professional organizations for architects and engineers have drawn up and published a standard section entitled "General Conditions" for use in the General Requirements division. These documents cover major topics similar to those listed above. Special items that pertain to a particular job are included in the Supplemental General Conditions that are written by the architect or engineer. Special laws or building code requirements are listed in the Supplemental General Conditions.

13.3 DIVISION 2 — SITE WORK

Division 2 describes work to be done on the site, including soil testing, core drilling, standard penetration tests, and seismic exploration. All of these tests must be conducted, and the requirements are expressed in this division.

The demolition of existing structures, materials, and debris is included in Division 2. The capping and removal of existing services is usually mentioned in this division even if the work is also included under the trade division. For example, the capping of an existing sewer line is mentioned in Division 2 and then is described in detail in Division 16 (plumbing).

In addition, the site work division describes the various steps in the clearing of the site. Clearing of the site includes structure moving, clearing and grubbing of shrubs, tree removal, pruning, and tree relocation. Also described are grading and earth-

work, grading and drainage, paving and surfacing, and landscaping.

Site improvements such as fences, gates, guardrails, lighting, and irrigation systems (site watering systems) are also included in this division.

13.4 DIVISION 3 — CONCRETE

All concrete work is described in this division. The major subdivisions included under Division 3 are concrete form work, expansion and contraction joints, cast-in-place concrete, specially placed concrete, and recast concrete.

13.5 DIVISION 4 — MASONRY

Division 4 of the specifications deals with brick, stone, fire brick, glass brick, clay backing tile, and ceramic veneer. Related topics such as joint reinforcement, mortar, anchors and tie systems, masonry joints, control joints, and masonry accessories are also included.

13.6 DIVISION 5 — METALS

The metals used in the building are specified under Division 5. If the building has a steel structure, the steel members are specified in this section. Metal roof decking, metal floor decking, and permanent metal forms are described in this section.

Metal fabrications such as stairs, handrails, railings, gratings, castings, and lintels are included in this division. In some cases, ornamental metals are used for these fabrications and are described in detail in this section.

13.7 DIVISION 6 — WOOD AND PLASTICS

Materials such as rough carpentry (framing), heavy timber construction, prefabricated structural wood, finished carpentry, and architectural woodwork (cabinets and built-in woodwork) are described in this division.

Plastics are used in many ways in modern construction. This section of the specifications is devoted to such items as joists, studs, columns, beams, hangers, connecting devices, and other miscellaneous plastic items not specifically identified under other headings.

13.8 DIVISION 7 — THERMAL AND MOISTURE PROTECTION

Divison 7 contains written descriptions of roofing materials, waterproofing materials, flashing and sheet metal trim, insulation, roof accessories, and sealants.

13.9 DIVISION 8 — DOORS AND WINDOWS

Metal doors and frames, wood and plastic doors, entrances and storefronts, metal windows, special windows, wood and plastic windows, hardware and specialties, and glazing are the main areas of concern in Division 8.

13.10 DIVISION 9 — FINISHES

Lath and plaster, gypsum wallboard, terrazzo, acoustical treatment, ceiling suspension systems, wood flooring finishes, carpeting, special flooring, floor treatment, painting, special coatings, and wall covering are found in this division of the specifications.

13.11 DIVISION 10 — SPECIALTIES

Division 10 of the specifications is a catch-all section. If items required for the construction do not fall under the other divisions of the specifications, they are usually found in Division 10.

Typical items in this division are chalkboards and tackboards, louvers and vents, grilles and screens, pest control, fireplaces, flagpoles, lockers, storage shelving, directional signage, and sun control devices.

13.12 DIVISION 11 — EQUIPMENT

Division 11 lists equipment which is normally furnished and installed by the general contractor as a part of the construction of the building.

Equipment described in this division includes maintenance equipment, bank and vault equipment, food service equipment, vending equipment, athletic equipment, laundry equipment, library equipment, medical equipment, waste handling equipment, and loading dock equipment.

13.13 DIVISION 12 — FURNISHINGS

Major headings in this division are artwork, window treatment, fabrics, furniture, rugs and mats, and furnishing accessories.

13.14 DIVISION 13 — SPECIAL CONSTRUCTION

The description and necessary information for constructing special areas are given in this division. Special construction areas include clean rooms, operating rooms in hospitals, incinerators, instrumentation rooms, nuclear reactors, radiation treatment rooms, sound and vibration rooms, vaults, and swimming pool spaces.

13.15 DIVISION 14 — CONVEYING SYSTEMS

Dumbwaiters, elevators, hoists and cranes, lifts, material handling systems, moving stairs and walks, and pneumatic tube systems are typical items described in Division 14.

13.16 DIVISION 15 — MECHANICAL

Division 15 covers information on air conditioning, ventilating and heating in a building. The mechanical division of the specifications generally has major sections consisting of general provisions (general description of the work, procedures, etc.) basic materials and methods, equipment, air distribution, piping, insulation, controls, and instrumentation.

> NOTE: Plumbing specifications are also included in the standard mechanical specification division; however, many engineers now write plumbing specifications as a separate division. With plumbing as Division 16, electrical becomes Division 17.

13.17 DIVISION 16 — PLUMBING

The plumbing division is similar to the mechanical section. Major sections consist of general provisions (general description of the work, procedures. and so forth), basic materials and methods, equipment, piping systems and services, and insulation. Division 16 also includes fire protection (sprinklers) and special services such as medical gases, fuel gases, compressed air, and process piping systems.

13.18 DIVISION 17 — ELECTRICAL

The electrical division usually consists of general provisions, basic materials and methods, power generation, power transmission, power service and distribution, lighting, special systems, communications, controls, and instrumentation.

13.19 RESPONSIBILITY AND LIABILITY

The responsibility for the design and all calculations for the construction project lies with the registered architect and registered engineer. (See Chapter 8 for information on the registration process for architects and engineers.) After satisfactorily fulfilling the registration requirements, architects and engineers are granted professional seals to be placed on all plans they are responsible for. Seals are also required on the title page of the specifications. (See Figure 14.3, page 78, for an example of the engineer's seal. The architect's seal is similar.)

With registration, the architect and the engineer assume, by law, responsibility for all work done by employees working under their guidance. In most cases these professionals carry liability insurance to protect them against possible financial ruin if a design failure should occur.

In the specifications, the architect and the engineer create a legal document with specific instructions to workers and construction technicians as to how the construction is to be performed. Specifications must be carefully written and followed to protect against poor workmanship, which could cause problems during and after construction.

SUMMARY

• Specifications contain information that is not shown on the plans.

• Except on very small jobs, specifications are usually printed and bound into book form.

• There are usually 16 divisions of information in a set of specifications (except when plumbing is separated from the mechanical division).

• Administrative, legal, and technical provisions are covered in Division l of the specifications.

• Divisions 2 through 17 include specific information pertaining to each area of construction on the job.

• The professional registered engineer and architect are responsible for the design and calculations on a construction project.

REVIEW QUESTIONS

1. Define specifications.

2. List five types of information that might be found on a typical set of specifications.

3. If a job is too small to necessitate printing the specifications in book form, where are the specifications placed?

4. Explain why specifications for government jobs are usually very long and complicated.

5. List three examples of information found under General Requirements.

6. What are "shop drawings"?

7. What information is listed under "Supplemental General Conditions"?

8. Give five examples of items or details that might be found under Division 2 — Site Work.

ADDITIONAL STUDENT EXERCISE: MATCHING SPECIFICATIONS

DIRECTIONS: Match each of the following Divisions of Specifications to the correct item from Answers A. through M.

Division 2 — Site Work   _____

Division 3 — Concrete   _____

Division 4 — Masonry   _____

Division 5 — Metals   _____

Division 6 — Wood and Plastics   _____

Division 7 — Thermal and Moisture Protection   _____

Division 8 — Doors and Windows   _____

Division 9 — Finishes   _____

Division 10 — Specialties   _____

Division 11 — Equipment   _____

Division 12 — Furnishings   _____

Division 13 — Special Construction   _____

Division 14 — Conveying Systems   _____

ANSWERS:

A. elevators

B. flagpoles

C. tree removal

D. artwork

E. brick work

F. entrances and storefronts

G. painting

H. vending equipment

I. roofing materials

J. recast concrete

K. finished carpentry

L. ornamental handrails

M. swimming pool areas

14

TITLE BLOCKS

OBJECTIVES

After studying this chapter, you should be able to:

- identify the title block on a set of plans, and explain the significance of information on it

- explain how sheets are titled and numbered

- tell when and how the revision of drawings takes place

- locate on the title block and discuss briefly the scales used for drawing the plans

- explain the significance of the seal(s) of the architect and engineer

- discuss reasons why there is usually a liability note in the title block

14.1 INTRODUCTION

The title block is an important part of every sheet in a set of drawings. Traditionally, the title block is located in the lower right-hand corner of each sheet of drawings (Figure 14.1). Some architects and engineers have chosen to locate the title block along the right-hand edge of the sheet, across the bottom of the sheet, or even across the top of the sheet. Since there is no standard rule describing the title block or its location, architects and engineers can place it wherever they wish (Figure 14.2). Regardless of the placement of the title block, the same information is shown.

A set of building plans consists of many sheets of drawings, each of which has a title block to identify the job, the title of the drawing, and other important information.

The title block is the same on each sheet (see set of plans at the end of this textbook). Most architects or engineers devise a standard title block and have it cut and printed on the drafting paper or film used by the firm. With this standard title block form already printed on each sheet, the drafter has only to fill in the information pertaining to that particular sheet. In this chapter the discussion will focus on the information shown in the title block.

14.2 THE NAME OF THE JOB

In the set of drawings for a construction project, each title block contains the name of the job. In some cases, the owner's name is included with the job name. The job address is also listed. Having the job name and the address of the project on the title block prevents confusion as to the project to which any sheet belongs.

In some cases, the architect or engineer works on more than one project for the same owner at the same time. For example, if the architect or engineer is preparing plans for a company that will have two different office buildings on adjoining sites, the job name will be the same; however, the street address will be different. Listing the street address identifies the drawings if they should become mixed.

BUILDING FOR JOE JONES			
321 EAST 32ND STREET			
SMALL CITY, U.S.A.			

| DRAWN BY:
JOHN DOE | **JOHN B. HART INC.**
ARCHITECTS & ENGINEERS
1520 PROFESSIONAL LANE
SMALL CITY, U.S.A. TEL: 800-321-1234 | DATE: |
| CHECKED BY:
J. B. HART | | SCALE:
1/8" = 1'-0" |

REVISIONS:					SHEET NO.
DATE:	DESCRIPTION:		BY:	NO.	**M-3**
					OF 6

Figure 14.1   Typical title block

Figure 14.2   Various locations of title blocks

14.3   THE ARCHITECT'S AND/OR ENGINEER'S NAME

The architect is usually the prime designer who is responsible for hiring engineers as consultants. In this case, the architect's title block is used on all plans for the project. If the job is an engineering job or the engineer is the prime designer, the title block is that of the engineer. It does not matter which title block is used (the architect's or the engineer's), but in most cases, only one type of title block is used for all drawings pertaining to a job.

Some architects and engineers have a small, stick-on title block that is put on the plans. If, for example, all drawings have been placed on drawing paper with the architect's title block, the engineer puts a small, stick-on title block with the name of the engineering firm, address, telephone number, and other pertinent information on sheets prepared by the engineer. This information identifies the firm that has done the engineering design work. If the engineer's title block is used, the architect puts a small, stick-on title block on the architectural portion of the drawings.

14.4   SHEET TITLE

Each sheet in a set of drawings has a sheet title. The title should represent the major item shown on that sheet. For example, the first floor plan may be drawn on the same sheet with several details and sections. Since the main purpose of the sheet is to show the floor plan, the sheet title reads "First Floor Plan." It is acceptable for the title to contain several items.

In the above example, the sheet title could also read "First Floor Plan, Sections, and Details."

14.5 NUMBERING THE SHEETS IN A SET OF DRAWINGS

The sheet number is an important item on the title block. Sheet numbers are usually broken down as follows: Architectural (A), Plumbing (P), Mechanical (M), and Electrical (E). (Mechanical is the term usually used for air conditioning and heating drawings.) Examples and explanations of sheet numbering are given below:

A-1 of 12: The architectural section has a total of twelve sheets. A-1 is the first of those twelve sheets.

P-1 of 3: The plumbing section has a total of three sheets numbered P-1, P-2, and P-3.

M-1 of 5: The mechanical section has a total of five sheets numbered M-1 through M-5.

E-1 of 8: The electrical section has eight sheets numbered E-1 through E-8.

This method of numbering sheets allows the reader to tell at a glance whether all sheets are in place, and if not, what is missing. (See the set of plans at the end of this textbook for other examples of numbering on construction plans.)

14.6 REVISIONS

Sometimes it is necessary to make changes on working drawings after they have been issued to contractors as construction plans. The changes are made on working drawings and are identified by the drafter so that the changes can be noted by the contractor (see again Figure 14.1). Each revision is numbered and dated in the title block of the working drawing. The working drawing is then reproduced and reissued to the contractor. The contractor gives the owner the estimated cost for making the revision to construction. If this price is agreed upon by the owner, the owner in turn gives approval, in writing, to the architect or engineer. The architect or engineer instructs the contractor, in writing, to proceed with the revision as shown on the revised plan. At this point, the revised plan becomes the construction plan. If the revision is not approved by the owner, the original construction drawing remains the construction plan for the building.

14.7 DATES ON WORKING DRAWINGS (CONSTRUCTION DRAWINGS)

The architect or engineer decides what date to put on the sheets. All sheets show the same date on the title block regardless of when they were drawn. This date is important to indicate that all sheets belong to that set of drawings. If the job must be redesigned because of changes requested by the owner, a new date is assigned to the new drawings, and the old drawings are disposed of. The date on each set of drawings identifies the set of drawings and eliminates possible confusion.

14.8 NAMES OR INITIALS FOUND IN THE TITLE BLOCK

Two names or initials appear on the title block. The drafter working on the plan signs or initials it where indicated. Usually a senior designer (or the registered architect or registered engineer) checks the drawing for accuracy before it is printed and initials or signs each title block as indicated. These two persons are responsible for the accuracy of the plan. Each initials or signs only the sheets for which he or she is responsible.

14.9 THE SCALE USED FOR THE DRAWING

There is a place on each title block where the drafter records the scale used for the drawing. Sometimes several drawings or details on the same sheet use different scales. If this occurs, a title is listed under each drawing or detail. The scale that was used for that individual drawing or detail is also listed under the drawing or detail.

In the space on the title block where the scale is normally shown, the phrase "as noted" is usually written. When "as noted" is placed in the scale block, the person reading the plans knows to look at each individual drawing for the scale used.

14.10 THE ARCHITECT'S AND ENGINEER'S SEAL

Architects and engineers must be registered professionals in a state to have the right by law to design buildings in that state. (See Chapters 8 and 13 for the major requirements for registration of architects and engineers.)

Registered architects and engineers have seals that are required on all drawings that they are responsible for. Figure 14.3 is an example of an engineer's seal. The architect's seal and the engineer's seal are similar. Both seals are usually placed beside the title block unless there is enough space within the confines of the title block.

Figure 14.3 Simulated typical engineer's seal

14.11 THE LIABILITY NOTE

The architect (or engineer) usually has the following note printed on the drawing paper or film used in the office:

"This drawing is the property of the architect. Written consent is required prior to reproducing, altering, or using the contents."

The architect or engineer who places the professional seal on the drawings indicates that a registered professional is responsible, by law, for the design and drawing of the plan. The liability notation clears the architect or engineer of any responsibility for the design shown on the drawings if this design, or any part of the design, is used on another job or by another architect or engineer or owner. In other words, the architect or engineer has been hired to design the job shown on the plans, and the unauthorized use of any part of the design anywhere other than on the job for which the plans were prepared releases the architect and engineer from liability.

SUMMARY

- A title block is found on each sheet of a set of plans.

- Information about the owner, job name, job address, and job number can be found on each sheet.

- Some sets of drawings use the architect's title block, and other sets of drawings use the engineer's title block.

- Sheet numbers indicate the page number and whether the drawing is part of the architectural, mechanical, or electrical plans.

- Revisions are numbered and dated, reproduced, and reissued to the contractor after the owner's permission to make revisions has been obtained.

- The same date is used on each sheet of a set of drawings regardless of when the actual drawing was done.

- The initials on the title block indicate what persons are responsible for accuracy of the drawing.

- If different scales are used on the same sheet, the scale for each drawing is indicated under that drawing.

- The liability note in the title block protects the architect or engineer in case of unauthorized use of a set of plans.

REVIEW QUESTIONS

1. Where may the title block be found on a set of drawings?

2. Why is it advantageous to have standard title blocks printed on a company's drawing paper or film?

3. Why does each title block contain the owner's name, job name, and job address?

4. Whose name is in the title block — the engineering firm or the architectural firm? Discuss briefly.

5. How is the title of a sheet determined?

6. Explain how sheets of a set of drawings are numbered for identification.

7. When and how are revisions to drawings made?

8. How is the date in the title block determined?

9. Whose initials are put in the title block?

10. If several scales are used on the same sheet, how is this information shown?

11. What is the purpose of the liability note in the title block?

PART II

FREEHAND SKETCHING AND DRAFTING

15

TYPES AND WEIGHTS OF LINES

OBJECTIVES

After studying this chapter, you should be able to:

- ■ identify the title block on a set of plans, and explain the significance of information on it

- ■ identify the most common types of lines drawn by drafters

- ■ distinguish between hard and soft graphite pencils, and tell when to use leads of varying degrees of hardness

- ■ distinguish between types of leads used on drafting films

- ■ demonstrate the ability to draw the types of lines discussed in this chapter

- ■ interpret a typical index of standard lines that might be used

15.1 INTRODUCTION

By varying the types of lines on a plan, the drafter helps the reader to distinguish between different kinds of information on the plan. Many types of lines are used on working drawings and in freehand sketching. Object lines, construction lines, extension lines, projection lines, cutting plane lines, dimension lines, hidden lines, center lines, break lines, and border lines will be discussed in this chapter.

The line weight that is selected for plans is very important. The weight of a line describes how dark and how wide the line is drawn. Usually, wider lines are darker than thinner lines. For ink drawings, the ink is usually the same color (conventionally, black), and all lines are the same so far as darkness is concerned. Line width for ink drawings is determined by the choice of pens used by the drafter. It is not uncommon to dilute the ink with water to produce a lighter line. This chapter will include a discussion of other techniques that are used to obtain different shades of lines drawn in ink. Lines drawn with graphite pencils or pencils designed to draw on plastic film sheets will also be discussed.

15.2 SCHEDULE OF LINES USED IN DRAFTING AND FREEHAND SKETCHING

Figure 15.1 is a schedule with examples of lines that are commonly used in drafting. These lines are drawn with standard drafting pens and black

	PEN NUMBER 0 .014 INCHES
	PEN NUMBER 1 .020 INCHES
	PEN NUMBER 2 .024 INCHES
	PEN NUMBER 3 .031 INCHES

Figure 15.1 Common line weights

drafting ink. Inking pens are marketed in sets consisting of several pens with different size points. A standard set of pens usually has Numbers 3x0, 00, 1, 2, 2$\frac{1}{2}$, 3, and 4. Numbers 6x0 and 4x0 are available in the smaller sizes, and Numbers 6 and 7 are available in the larger sizes. Each pen draws a line of a specific width. The pen size is given in Figure 15.1.

Figure 15.2 shows a typical line schedule, or alphabet, for drawing plans. Each line has two characteristics, width and darkness. Some lines are drawn continuously, some are a series of short

SCHEDULE (ALPHABET) OF LINES		
NAME OF LINE	EXAMPLE OF LINE	DESCRIPTION OR USE
CONSTRUCTION LINE	THIN	ESTABLISHES SHAPES PRIOR TO DRAWING THE SAME LINES AS OBJECT LINES
OBJECT LINE	THICK	HEAVY LINE USED TO DRAW OBJECTS
BROKEN LINE	MEDIUM	USED TO SHOW PARTS OF OBJECT THAT ARE NOT VISIBLE
CENTER LINE	THIN	SHOWS CENTER OF OBJECT
EXTENSION LINE	EXTENSION LINE THIN	LINES EXTENDED FROM OBJECT SO DIMENSION CAN BE SHOWN
DIMENSION LINES	2' 0" 2' 0" 2' 0" THIN	LINE DRAWN BETWEEN EXTENSION LINES WITH DISTANCES
LEADER	OR THIN	LIGHT LINE (WITH ARROW HEAD DRAWN BETWEEN NOTES AND WHERE THE NOTE APPLIES)
SHORT BREAK LINE	MEDIUM	INDICATES SHORT BREAKS
LONG BREAK LINE	MEDIUM	INDICATES LONG BREAKS
CUTTING PLANE	A 1 OR THICK	SHOWS INTERNAL FEATURES, SECTION SHOWN LOOKING IN DIRECTION OF ARROW
PROPERTY LINE	THICK	SHOWS EXTENT OF PROPERTY
BORDER LINE	THICK	USED TO SHOW BORDER ON PLANS
NEW CONTOUR LINE	60 THIN	LINE INDICATES LAND ELEVATION. NUMBER INDICATES FEET ABOVE SEA LEVEL
EXISTING CONTOUR LINE	60 THIN	LINE INDICATES LAND ELEVATION. NUMBER INDICATES FEET ABOVE SEA LEVEL
PROJECTION LINE	THIN	LINE USED TO EXTEND DIMENSIONS, LOCATION POINTS FROM ONE SPACE TO ANOTHER

Figure 15.2   Typical alphabet of lines (Line Schedule)

dashes, and some are a series of short dashes and dots and long dashes and dots. Some are drawn with wide pen points and some are drawn with narrow pen points. Each line is used for a particular purpose. It is very important for the drafter to use the proper lines to relay the information needed and for these lines to be uniform. In other words, all object lines should be the same weight throughout the drawing, and all dimension lines should be uniform throughout the drawing.

15.3 LINES DRAWN WITH PENCILS

Lines drawn with pencils and those drawn with inking pens follow the same general rules. The weight of the lines drawn by a pencil can be controlled by the sharpness of the pencil and by the softness of the lead. If a heavy line is required, the pencil point must be dull so that the resulting line will be heavy. On the other hand, if the pencil point is sharp, the resulting line will be thin. Pencil-drawn lines should be about the same weight and darkness as inked lines (see Figure 15.1, page 83). Of course, pencil lines will not be as dark as inked lines; however, when a print is made from the penciled drawing, the print should reflect good, firm lines.

There are standard types of graphite pencils and special leads used with drafting films such as Mylar® or Vellum®. Graphite leads progress from B, which is very soft, up through HB, H, H-2, H-3, H-4, and so on. The higher the number, the harder the lead. The average drafter finds that the H and H-2 are in a good range of hardness. This lead is used for drawings and for lettering. Harder leads such as H-4 are sometimes used as guidelines or construction lines. As these lines are drawn lightly on the plan, a sharp pencil point must be maintained so that the lines do not show up on prints. The lines also can be easily erased when the drawing is finished.

Graphite leads and film leads are available in wooden pencils, and singular leads are available for mechanical pencils. Wooden pencils with film lead start with E-0 and progress upward through E-1, E-2, E-3, E-4, and so on. Each higher number represents a harder lead. Some mechanical pencil leads are numbered in a "P" series with P-1 the softest, and P-2, P-3, P-4, and so forth becoming progressively harder. In most cases the "P" numbers and the "E" numbers are identical in hardness.

Drafters who draw lines with pencils develop techniques to make all the lines on the drawing more uniform and the different weight lines more evident. For example, drafters learn that if the pencil is rotated as the line is being drawn, the line is more uniform from one end to the other.

15.4 OBJECT LINES

Object lines are used to outline the object being drawn. These lines should be drawn heavy (thick) and solid. These lines should be so distinct that there is no question as to the outline of the object being drawn (Figure 15.2).

15.5 HIDDEN LINES

An object on a drawing must be complete and, in most cases, must contain lines to represent all edges and interior sections and surfaces in the object. These edges and interior sections are shown with hidden lines. Hidden lines are usually broken lines that are thinner than object lines (Figure 15.3).

15.6 EXTENSION AND DIMENSION LINES

Extension and dimension lines are used to show dimensions of an object. Extension lines are used to

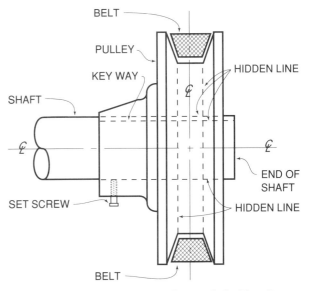

Figure 15.3 Belt-driven pulley with hidden lines

show the limits of the dimension given. These lines are thin, solid lines that are projected from the object at the exact place where the dimension is given. There is a small space between the end of the extension line and the object line. The dimension line is drawn between the two extension lines with arrowheads on each end where the dimension line and the extension line touch (Figure 15.4). Extension and dimension lines are thin, solid lines that should be the same weight.

Extension and dimension lines are very important because the size of the object is reflected by these dimensions. A mistake in a dimension line can create many problems in construction.

15.7   CONSTRUCTION LINES

Construction lines are used to help develop the drawing. Construction lines are thin lines that are drawn lightly on the plan (Figure 15.2). If object lines are put on the drawing while the object is being developed, mistakes are hard to erase. If construction lines are used instead, they are easily erased and changed. When the object has been fully developed, object lines can be drawn over the construction lines for the finished drawing. Unused construction lines are erased when the drawing has been completed. See Figure 15.5 for an example of an object being developed, first with construction lines, and then with object lines applied over the construction lines later.

Figure 15.4   Extension and dimension lines

Figure 15.5   Object and construction lines

15.8   PROJECTION LINES

Projection lines are like construction lines, though these fine lines may be drawn either heavy or light depending on how the lines are to be used. In orthographic projection (Chapter 16), projection lines are used to project dimensions from one view to the other (Figure 15.6). Sometimes these lines are left on the drawing, and sometimes the lines are erased when the drawings have been completed. In either case, projection lines are fine lines used in drafting to aid the drafter who needs to project dimensions from one area of a drawing to another.

15.9   CENTER LINES

It is important to show the center line of some objects or drawings so that dimensions can be referenced to the center of the drawing. The center

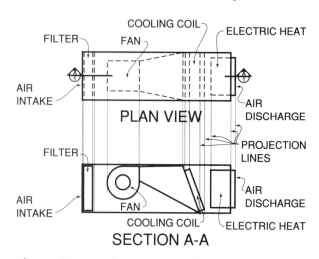

Figure 15.6   Using projection lines

line is usually shown as a series of long lines and short lines (Figure 15.2, page 84). The center line is a sharp, thin line that is drawn heavy enough to print along with the object line.

Two center lines are usually drawn at 90 degrees to each other so that the exact center is indicated where the two lines cross (Figure 15.7).

15.10   CUTTING PLANE LINES

Cutting plane lines can be shown with several types of symbols. One of the most common is a heavy line with a long dash and two short dashes drawn in series. The line is drawn through the object and extended for a short distance outside the object line. At this point a short line is drawn showing the direction in which the cutting plane is cut, and an arrowhead is placed on the end of the short line (Figure 15.2). The cutting plane is numbered to identify the section if it is located somewhere else on the plan.

A second method of identifying a cutting plane is to draw a heavy line consisting of long dashes through the object. The line is extended to the outside of the object, and a short line with arrowheads is added similar to the cutting plane described above. The cutting plane is identified with a number or letter so that the section can be identified when it is drawn on the sheet. Refer to Figure 10.7, page 54 in Chapter 10 for additional examples of the use of cutting plane lines.

One of the most popular cutting plane symbols used in architectural and mechanical drawings is a heavy line drawn from a point inside the object to a point outside the object on both sides of the object or where the cutting plane ends. A continuous line

between these two locations is not necessary. A circle is drawn at each end of the cutting plane, and the circle is cut in half with a horizontal line. An arrowhead is created around the circle to show the direction the section is shown. The top half of the circle contains the section number. The bottom half of the circle gives the sheet number where the section is shown (Figures 15.2 and 15.6).

15.11   BREAK LINES

It is sometimes necessary to increase the scale of details of various parts of a building. With the larger scale, it is not practical or necessary to draw the entire portion of the building. When this is the case, it is desirable to show what is needed and omit the unnecessary portion of the building. Break lines are used to show that there is a continuation of the detail; however, the continued drawing is not necessary. Figure 15.2 demonstrates two types of lines used as break lines. The wavy line is drawn freehand, and the ruled line and zig-zags are drawn with drafting instruments. As a general rule, the wavy, freehand line is used when the break is short, and the ruled line with zig-zags is used when the break is long (Figure 15.5).

15.12   BORDER LINES

Border lines are used to frame the drawing sheet. Horizontal border lines are drawn across the top and bottom of the sheet, and vertical border lines are drawn on the left side and right side of the sheet. The border lines set the size of the drawing and are containment lines for the drawing (or drawings) on that sheet. Border lines are usually very heavy, wide lines that are two or three times the weight of object lines (Figure 15.2).

15.13   SCHEDULE OF LINES
USED IN OFFICES

Many types of lines are used by drafters in various offices. An office often encourages uniformity by using a line schedule to show that certain lines

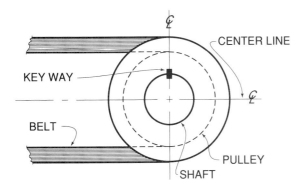

Figure 15.7   Pulley and shaft showing center lines

stand for specific things on drawings made in that office. A typical alphabet, or schedule, of standard lines for one office is included in Figure 15.2. As a rule, standard lines are identified and approved by the office manager for use by drafters in that office.

SUMMARY

• Line weight is determined by the size of the pen used by the drafter.

• Line weights drawn by pencil are controlled by the sharpness of the pencil and quality of the lead.

• Object lines are heavy solid lines that outline the object being drawn.

• Interior sections are shown by thin broken lines called hidden lines.

• Extension and dimension lines show the dimensions of an object.

• Construction lines are light, thin lines that are usually erased when the drawing is completed.

• Projection lines project dimensions from one view to another.

• Center lines show the exact center of an object.

• Various symbols indicate cutting plane lines.

• Break lines are used to indicate that only a portion of an object is being drawn.

• Border lines frame the drawing sheet on the top, bottom, left side, and right side.

• Some offices have their own schedule or alphabet of standard lines to be used in that office.

REVIEW QUESTIONS

1. Why does a drafter need more than one inking pen to draw a plan?

2. How does the drafter control the weight of lines drawn by pencil?

3. What is the most common lead used by drafters using standard graphite pencils?

4. Why should the drafter learn to rotate the pencil as lines are being drawn?

5. Why is it especially important that dimension and extension lines be accurate?

6. What is the purpose of construction lines, and why are they usually erased after the drawing has been completed?

7. Demonstrate one method of drawing cutting plane lines, using an example from the textbook.

8. How does the use of break lines save time for the drafter?

9. Why do many offices have an index of lines for that office?

16

ORTHOGRAPHIC PROJECTION

OBJECTIVES

After studying this chapter, you should be able to:

- explain why orthographic projection is used for drawings

- define a viewing plan

- describe the three basic views used in orthographic projection

- describe two other views, other than the three basic views, that are often used in drawing plans

- describe the various types of lines used in orthographic projection and give their functions

- be able to distinguish between the views used in drawing plans

- explain how the various views of building plans correspond to the five views used in orthographic projection.

16.1   INTRODUCTION

In this unit, orthographic projection will be discussed and illustrated. In order to relay information clearly from designer to fabricator or craftsman, it is important to show graphically the shape of an object, the dimensions of the object, and the relationship of one view to other views of the object. Orthographic projection is one of the most common methods for drawing plans of an object to relay important information about it.

Orthographic projection is often referred to as making three-view drawings of an object. In most cases, three views are enough to relay the needed information to the fabricator or craftsman. The three basic views are the *plan view*, *front view*, and *end view*.

Although the three basic views are usually sufficient, a total of five views can be developed by using projection lines from one view to the other. The five views that can be developed are the plan view, the front view, the right side or end view, the left side or end view, and the rear view. The plan view can be a floor plan or roof plan. The front view is usually called the front elevation, the right and left side views are called the right and left elevations, and the rear view is called the rear elevation.

16.2   THE VIEWING PLANE

Understanding orthographic projection requires an understanding of the viewing plane. A plane is an imaginary, flat, transparent surface similar to a pane of glass. When this plane is placed parallel to the surface of an object, the surface of the object parallel to the plane is shown in its true size and shape. If the surface of the object is not parallel to the plane, the surface is shown as it is seen according to size and shape (Figure 16.1).

When surfaces are not parallel to the plane, the orthographic projection view shows these surfaces smaller in width or height (whichever the case may be). See Figure 16.2 for examples of object surfaces that are not parallel to the viewing plane.

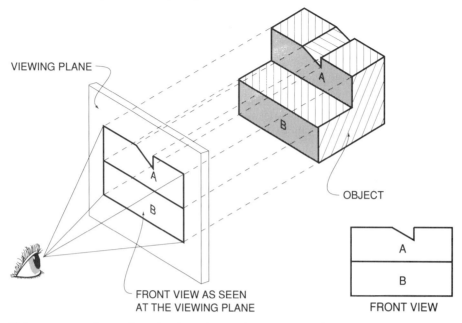

Figure 16.1 Object as seen through a viewing plane—front view

For the plan view, the viewing plane is parallel to the top of the object (Figure 16.2). The shaded portion represents the surface of the object that is

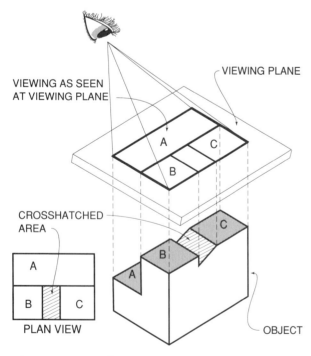

Figure 16.2 Object as seen through a viewing plane—plan view

parallel to the viewing plane. The crosshatched portion represents the surface of the object that is not parallel to the viewing plane, and this surface is not shown actual size. (Surfaces that are not parallel to the viewing plane are shown as they appear rather than actual size.)

The front view of the object is shown in Figure 16.1 along with the viewing plane. The viewing plane is vertical, and surfaces of the object that are parallel to the viewing plane are shown shaded. Surfaces that are not parallel to the viewing plane are shown crosshatched. As in the plan view, surfaces that are not parallel to the viewing plane are shown as they are seen and do not reflect true shape and size.

The right side view is shown in Figure 16.3. The viewing plane is shown in the vertical position, and the surfaces that are parallel to the viewing plane are shown shaded. Surfaces that are not parallel to the viewing plane are shown crosshatched. Again, surfaces that are not parallel to the viewing plane are shown as they are seen and do not reflect true shape and size.

The left side view can be shown in the same manner as the right side view, and the rear view (back view) can be shown in the same manner as the front view, with the viewing plane in the vertical position at the rear of the object.

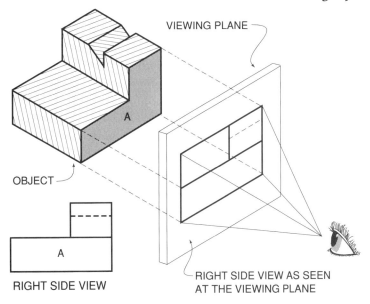

VIEWING PLANE

OBJECT

A

RIGHT SIDE VIEW

RIGHT SIDE VIEW AS SEEN
AT THE VIEWING PLANE

Figure 16.3   Object as seen through a viewing plane—right side view

16.3   COMBINING VIEWING PLANES

If viewing planes were placed on all four sides of an object and on top of the object, the object would be encased in a transparent box. If the image of the object, as seen through the various viewing planes, were projected to each side of the box, the plan view, front view, rear view, right side view and left side view would be visible on the viewing planes. In Figure 16.4, the plan view is projected straight down. The other views are aligned with the plan view. When all of the various views are considered, the true lengths, widths, and heights are apparent.

16.4   TYPES OF LINES USED IN ORTHOGRAPHIC PROJECTION

Several types of lines are used in orthographic projection drawings. Described briefly in this unit will be object lines, construction lines, hidden lines, miter lines, projection lines, extension lines, and dimension lines. A more detailed discussion of types of lines used in drafting was included in Chapter 15.

Object lines are drawn for the purpose of outlining the object. Object lines are moderately heavy and dark because they show the shape of the object (Figure 16.5). A 2H pencil is normally used to

draw the object lines. Refer to Chapter 15 for a more detailed description of types of pencils and drawing pens, line weights, and drawing techniques.

Construction lines (often called projection lines) are usually drawn with a harder pencil (H4 or H6). These lines are thin and are usually not as dark as object lines. Construction lines are used to transfer distances from one view to the other in orthographic projection (see Figure 16.5). Construction lines are usually removed from the finished drawing.

Hidden lines are broken lines that are used to outline surfaces that cannot be seen in the other views because the surfaces are located behind another surface. Hidden lines are usually drawn with the same thickness and weight used for regular object lines. The H2 pencil is usually used to draw hidden lines (see Figure 16.5).

Miter lines are construction lines drawn at 45 degrees to the horizontal of the drawing. Miter lines change the direction of distances when construction lines are drawn from one view to the other (see Figure 16.5). When projection lines are extended from the object in one view to the miter line and then are drawn up or down from the miter line, the true size or dimension of that part of the object is transposed to the other view.

Projection lines are a type of construction line, and they are usually removed from the finished drawing. Projection lines, as described in the pre-

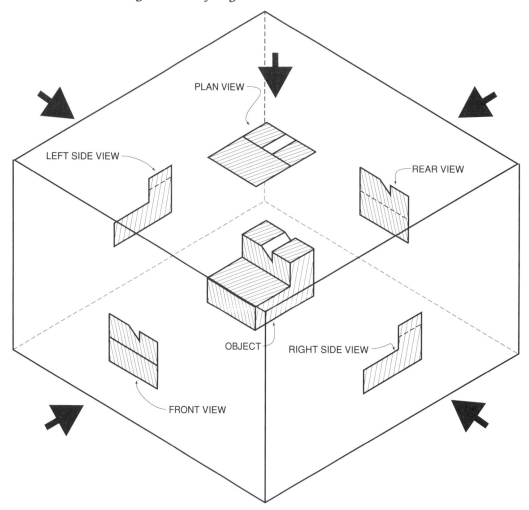

Figure 16.4 Plan view, front view, right side view, and rear view with five viewing planes

ceding paragraph on miter lines, are drawn from the object to the miter line. By projecting these lines up or down on the drawing, actual distances or dimensions can be transferred from one view of an object to another. This procedure aids the drafter, who does not have to measure each dimension on each view of the object (Figure 16.5).

Extension lines are construction lines that are drawn a little heavier than projection lines and are left on the finished drawing. These lines are extended out from the object to reflect the distance between various points on the object. Extension lines do not touch the object, and they are used to indicate where measurements of the object start and stop (see Figure 16.8, page 96).

Dimension lines are construction lines drawn like the extension lines described above. These lines are

drawn between extension lines and have arrowheads on each end of the line. This is the usual method for showing dimensions of the object. The dimension or measurement is written on the dimension line. Dimensions are usually given in feet and inches. In some cases, dimensions are given in decimal form or in metric form.

16.5 USING ORTHOGRAPHIC PROJECTION TO CONSTRUCT THE THREE BASIC VIEWS

In Figures 16.6, 16.7, and 16.8, pages 94, 95, and 96, the object is shown isometrically in the upper

Figure 16.5 Orthographic projection

left-hand corner of the drawing. The orthographic method of drafting is used to show the three basic views (plan view, front view, and right side view) of the object.

Figure 16.6 is an example of the plan view of a building. Projection lines have been extended to the right and to the bottom from the major points on the plan view.

The front view can be developed using projection lines from the plan view and the known dimensions of the height of the object (Figure 16.7). Projection lines are then extended to the right from the main points of the front view. A miter line is drawn through the projection lines extended to the right from the plan view. The projection lines are

extended down from the miter line to a point where they intersect the projection lines from the front view (Figure 16.7). The projection lines extended down from the miter line now intersect the projection lines from the front view. By darkening the projection lines between these intersections, the right side view can be drawn without having to do any measuring (Figure 16.8).

Surfaces that cannot be seen from the imaginary viewing plane are shown with broken lines (hidden lines) as illustrated in Figure 16.8. The blueprint reader can tell that there is something there, even if the hidden detail cannot be seen. By looking at all three drawings, the reader can determine what the hidden lines are showing.

The location of the miter line drawn through the horizontal projection lines determines the location of the right side view (Figures 16.6–16.8). Before drawing the miter line, the drafter must be sure that the right side view will be located where one view can be related easily to the other.

When the drawing is complete, the drafter can add extension lines and dimension lines as shown on the plan view and front view to give exact dimensions of the object (Figure 16.8, page 96). One dimension of Figure 16.8 is shown on the right side view; however, this dimension could also be given on the front view.

16.6 SELECTING THE PROPER VIEWS

When an orthographic projection drawing is constructed, it is important that the views that are selected show the most detail of the object being drawn. Once the front view and the right side view are selected, the plan view can be drawn, and the front view and right side view can be developed. It is sometimes necessary to draw the front view, right side view, left side view, and rear view to show all the necessary details of the object. If the front view and the right side view do not do an adequate job of showing necessary details, then the left side and rear views must be shown.

It is always important to locate the drawings on a sheet so that the entire drawing can be shown and so that each view relates to the other views.

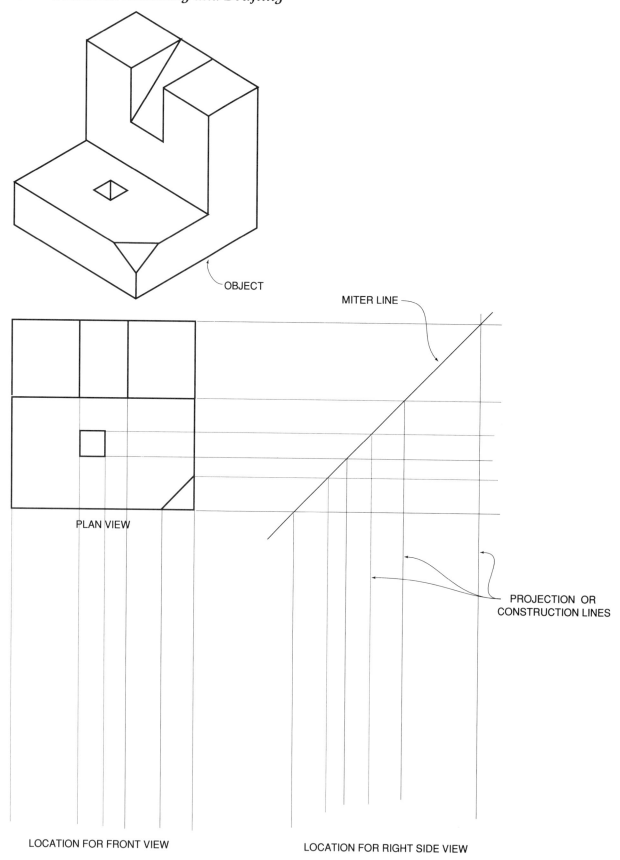

OBJECT

MITER LINE

PLAN VIEW

PROJECTION OR
CONSTRUCTION LINES

LOCATION FOR FRONT VIEW

LOCATION FOR RIGHT SIDE VIEW

Figure 16.6   Orthographic projection showing miter line and projection lines from plan view

OBJECT

MITER LINE

PLAN VIEW

PROJECTION OR
CONSTRUCTION LINES

PROJECTION
OR
CONSTRUCTION
LINES

FRONT VIEW

LOCATION FOR RIGHT SIDE VIEW

Figure 16.7 Orthographic projection showing miter line and projection lines from plan view and front view

Figure 16.8 Orthographic projection showing plan view, front view, and right side view with projection or construction lines, and extension and dimension lines

16.7 WORKING DRAWINGS AND ORTHOGRAPHIC PROJECTION

Architectural and engineering plans for a building use the orthographic projection procedure on the working drawings. A comparison of building plans with the orthographic projection procedure is as follows:

1. The floor plans and the roof plan represent the plan view of the building.

2. The front elevation, the right side elevation, left side elevation, and the rear elevation plans represent the front view, right side view, left side view, and back view of the building.

SUMMARY

• Orthographic projection refers to the making of three-view drawings of an object.

• The three basic views of an object are the plan view, front view, and end view (right side view).

• The five views which can be created in orthographic projection are the plan view, front view, right side or end view, left side or end view, and rear view.

• A viewing plane is an imaginary flat, transparent surface through which the drafter views an object.

• Surfaces that are parallel to the viewing plane are shown in true shape and size.

• Surfaces not parallel to the viewing plane are not shown in their true width or height.

• Lines used in orthographic projection include object lines, construction lines, hidden lines, miter lines, projection lines, extension lines, and dimension lines.

• Orthographic projection views should be carefully selected to show the most important details.

REVIEW QUESTIONS

1. Explain briefly why orthographic projection is used on plans by architects and engineers.

2. Describe the three basic views of an object.

3. Define a viewing plane.

4. An object is drawn in its true size and shape if

   (A) the surface of the object is parallel to the viewing plane.

   (B) the surface of the object is not parallel to the viewing plane.

5. Define each type of line and tell how it is used:

   (A) object line

   (B) construction line

   (C) hidden line

   (D) miter line

   (E) projection line

   (F) extension line

   (G) dimension line

6. Floor plans of a building represent the (plan view, front view) of a building.

7. Building plans of the front elevation represent the _____ view in orthographic projection.

8. The rear elevation is the same as the _____ .

ADDITIONAL STUDENT EXERCISES: Completing Orthographic Projection Views

For Problems 1 through 5, the student should draw the missing lines needed to complete the three basic orthographic projection views of the indicated isometric object.

Figure 16.9 Exercise 1

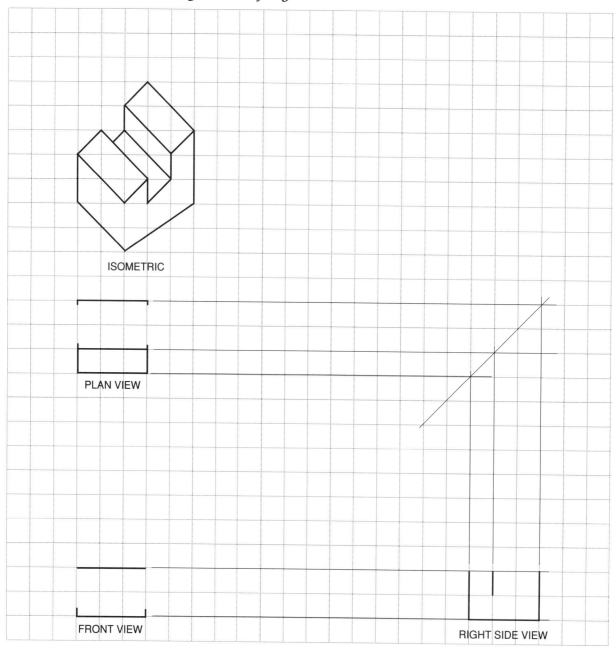

ISOMETRIC

PLAN VIEW

FRONT VIEW

RIGHT SIDE VIEW

Figure 16.10   Exercise 2

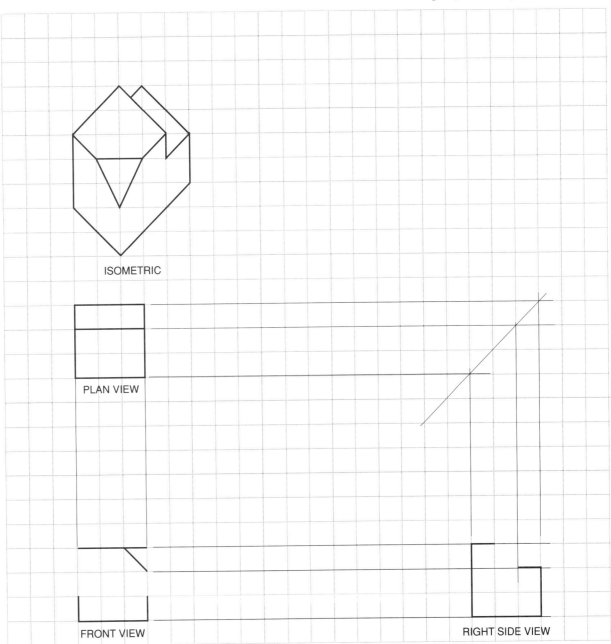

ISOMETRIC

PLAN VIEW

FRONT VIEW

RIGHT SIDE VIEW

Figure 16.11 Exercise 3

Figure 16.12   Exercise 4

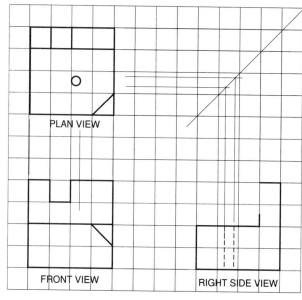

Figure 16.13   Exercise 5

17

OBLIQUE DRAWINGS

OBJECTIVES

After studying this chapter, you should be able to:

- describe three types of drawings that show three-dimensional views of objects

- explain how construction lines can be drawn to show the three basic views of an object

- demonstrate the ability to create an oblique drawing of an object

- explain how variations of oblique drawings are made

17.1 INTRODUCTION

Information can sometimes be relayed better when the object being drawn takes on the feeling of a three-dimensional rather than a two-dimensional drawing. A two-dimensional drawing shows width and height whereas a three-dimensional object has width, height, and depth. An oblique drawing is one type of pictorial drawing that relays this three-dimensional feeling. The oblique drawing is very helpful for showing details of air-conditioning equipment.

Normally, objects are viewed in three dimensions. The plan view, front view, and side view are seen all at once. The three basic views of an object are discussed in Chapter 16.

There are several ways to draw an object so that it appears to be three-dimensional. One method is to make isometric drawings. (Isometric drawings are discussed in Chapter 18.) A second method is to make perspective drawings with diminishing points. (Perspective drawings are not included in this book because they are not usually used in the heating and air-conditioning fields.) Oblique drawings are probably the easiest of the drawings that give a three-dimensional impression. Oblique drawings show two or more surfaces at one time on the same drawing. The front view appears in its true size and shape, but the other sides are not shown in their true size and shape because they are drawn at an angle.

17.2 HOW OBLIQUE DRAWINGS ARE CREATED

The orthographic projection method of drafting is discussed in Chapter 16. Using orthographic projection, the drafter draws an object showing the three basic views (the plan view, front view, and right side view). The drawings are not drawn to scale.

Oblique drawings consist of a front view or an end view of an object drawn in the same manner as in orthographic projection. This view is drawn with the surface parallel to the imaginary vertical viewing plane. (See Chapter 16 for an explanation of the viewing plane.) Figure 17.1 shows the basic

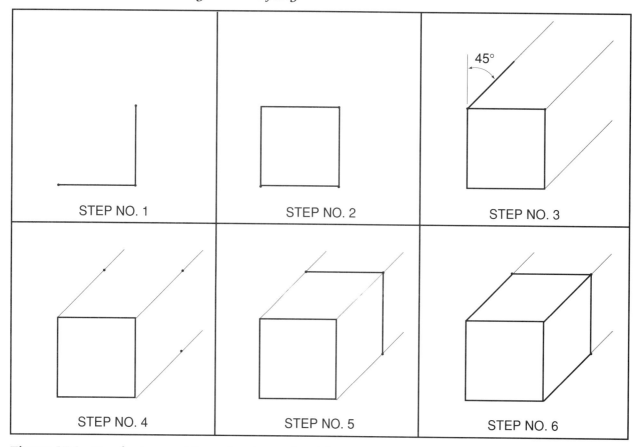

STEP NO. 1   STEP NO. 2   STEP NO. 3

45°

STEP NO. 4   STEP NO. 5   STEP NO. 6

Figure 17.1   Developing an oblique drawing to the right

steps in making an oblique drawing. Step 1 demonstrates how the front view is started, and Step 2 shows how the front view is completed.

When the front view or side view has been established, the drafter projects construction lines to the upper right or upper left from each corner or change in direction of planes of the surfaces of the object. These construction lines are projected back from the front or side view of the object at 45 degrees from the horizontal of the drawing (see Step 3). When all of the projected construction lines have been drawn, the drafter locates points along these lines to establish distances from the front or side view, thus revealing the back outline of the object (see Step 4). The points should be placed in such a manner that the drawing will be in proportion. In other words, the distance along the projected construction lines should be representative of the distance between the front of the object and the back of the object.

When these points have been established and the distances verified, the drafter connects the points with vertical and horizontal lines to form the out-

line of the back of the object (see Step 5). Step 6 shows the finished drawing of the object. Angular lines can be drawn where needed to show inclined planes in the object.

NOTE: In this discussion, consider the back (or rear) view as the side that is opposite the front or end view (the view that was first drawn).

Intermediate points can now be placed on the projected construction lines where needed so that object lines can be drawn between these points to indicate changes in direction of planes on the object.

The only views of the object that are not shown in these two drawings are the back view and the bottom view. If these views are required, the drafter can rotate the object and use the oblique method of drafting to show these sides or views.

Figure 17.2 shows the drawing process for extending the construction lines to the left. The steps are the same as those described in Figure 17.1 except that the construction lines are extended to the left instead of the right.

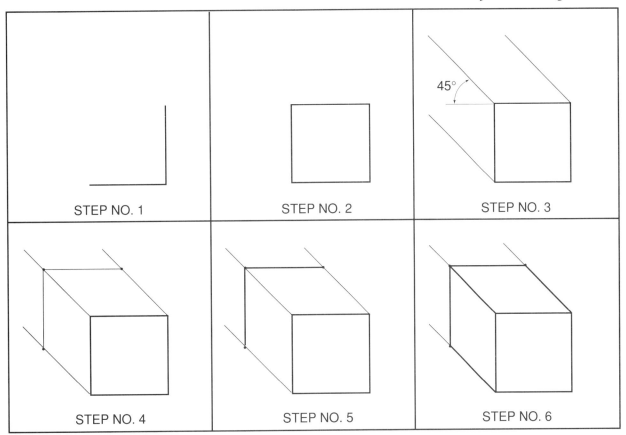

Figure 17.2 Developing an oblique drawing to the left

17.3 USING THE ARCHITECT'S SCALE FOR OBLIQUE DRAWINGS

The architect's scale is a valuable tool in freehand sketching because it helps establish distances and helps keep the drawing in proportion. If a scale drawing is needed, the architect's scale is used for the entire drawing.

The oblique drawing is not intended to be a scale drawing. Even though many of the dimensions are drawn to scale, the oblique drawing is not a scale drawing because some of the surfaces appear distorted if they are at an angle to the viewing plane. The purpose of an oblique drawing is to give a pictorial view or three-dimensional concept of an object.

17.4 LINES DRAWN ON OBLIQUE DRAWINGS

Lines drawn on freehand sketches and on instrument drawings should be drawn as construction lines in the beginning. After the object has been constructed with these construction lines, object lines can be drawn over the construction lines to show the object clearly. When this step has been completed, the construction lines should be removed to make the drawing neater.

When the oblique drawing is completed, the object will have a three-dimensional appearance. In the drawing, the object is shown as it actually looks from three sides: the front view (if this is the view that the drafter started with), the side view, and the plan view. (See Step 6 of Figure 17.1.) The oblique drawing is helpful because it reveals the three basic viewing planes, and it is drawn on a single plain piece of paper.

17.5 VARIATIONS IN OBLIQUE DRAWINGS

In some cases, drawings may appear to be distorted. This can be corrected by adjusting the lines as necessary to obtain realism (Figure 17.3). This adjustment can be made by shortening or lengthening lines.

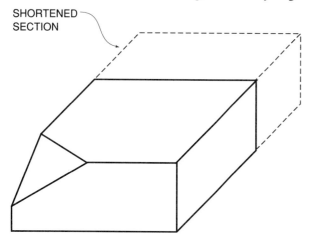

SHORTENED SECTION

Figure 17.3 Adjusting drawing to obtain realism

Oblique drawings can be drawn at angles other than 45 degrees. In some cases it is more desirable to project construction lines at 30 degrees or 60 degrees. Figure 17.4 shows an oblique drawing of a heating coil with construction lines drawn at 45 degrees. Figure 17.5 shows the same heating coil with construction lines drawn at 30 degrees. Figure 17.6 shows the same coil with the construction lines drawn at 60 degrees. The angle chosen for the construction lines makes a slight difference in appearance of the three drawings.

30°

HEATING COIL

Figure 17.5 Drawing of a heating coil with projection at 30 degrees

SUMMARY

• An oblique drawing is a type of pictorial drawing that gives a three-dimensional effect.

• Isometric drawings and perspective drawings also give a three-dimensional effect.

• Oblique drawings are not scale drawings.

• Construction lines project back from the corners of the front view in order to establish the back outline of the object.

45°

HEATING COIL

Figure 17.4 Drawing of a heating coil with projection at 45 degrees

CONSTRUCTION LINES

60°

HEATING COIL

CONSTRUCTION LINES

Figure 17.6 Oblique drawing of a heating coil (with construction lines) with projection at 60 degrees

- An oblique drawing reveals the three basic viewing planes of the object (front view, side view, and top view).

- By rotating the object, the drafter can also show the other side view, the back view, and the bottom view.

REVIEW QUESTIONS

1. Describe an oblique drawing.

2. List three types of drawings that give a three-dimensional effect.

3. What are the three basic views of an object?

4. After the front or side view has been drawn, lines are projected back from the corners of the object at how many degrees?

5. How is the back outline established?

6. Is an oblique drawing a scale drawing?

7. How is it possible to use an oblique drawing to show the bottom view of an object?

ADDITIONAL STUDENT EXERCISES: CONSTRUCTING OBLIQUE DRAWINGS

1. Draw an oblique drawing from the orthographic drawings in Figure 17.7.

2. Draw an oblique drawing from the orthographic drawings in Figure 17.8.

3. Using the elevation views shown in Figure 17.9, project the necessary lines to show the duct fitting as oblique drawings.

Figure 17.7 Exercise 1

Figure 17.8 Exercise 2

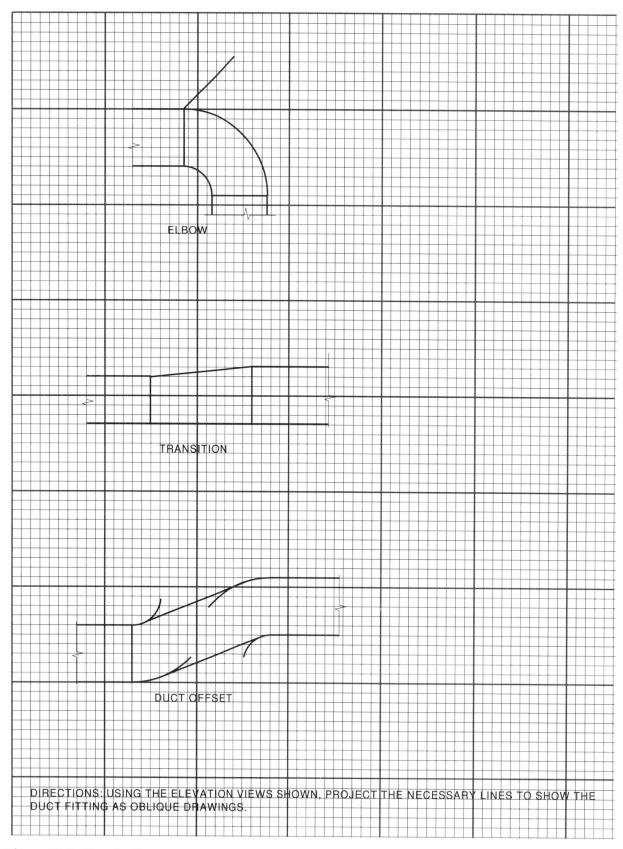

ELBOW

TRANSITION

DUCT OFFSET

DIRECTIONS: USING THE ELEVATION VIEWS SHOWN, PROJECT THE NECESSARY LINES TO SHOW THE DUCT FITTING AS OBLIQUE DRAWINGS.

Figure 17.9   Exercise 3

18

ISOMETRIC DRAWINGS

OBJECTIVES

After studying this chapter, you should be able to:

- identify the steps used to create the three basic views of an object isometrically

- demonstrate the ability to construct isometric drawings of simple objects

- discuss reasons why isometric drawings are the most commonly used pictorial drawings in construction projects

- compare isometric drawings drawn at 30 degrees from the horizontal with those drawn at 60 degrees from the horizontal

- explain the advantages of using isometric drawings for piping diagrams

18.1 INTRODUCTION

Chapter 17, "Oblique Drawings," discusses the use of pictorial drawings to relay information from the drafter/designer to the craftsman or construction worker. The isometric drawing method is also used to give the reader of the drawing a feeling of seeing objects in three dimensions.

As stated in Chapter 17, the *easiest* of the pictorial drafting methods is the oblique drawing method; however, the isometric drawing method is more commonly used because it is *easier to read*. If several objects are shown in the same drawing, there appears to be more space between the objects in an isometric drawing.

A third type of pictorial representation is the perspective drawing with diminishing points. Since this drawing method is very seldom used in the HVAC field, perspective drawings are not included in this book.

In this chapter, we will discuss the development of an isometric in steps. As stated in Chapter 16, the three basic views can be the plan (top), front, and left (or right) side views. This is also true with isometric drawings. The sides are determined by the drafter. The drafter can rotate the object to reveal the most information on the front and right-hand side views.

Figures 18.1 through 18.9 are step-by-step drawings that illustrate how a drawing is developed. Figure 18.10 shows the final (finished) drawing of an object.

18.2 HOW ISOMETRIC DRAWINGS ARE DRAWN

Isometric drawings show three basic views of an object. These three views are the plan view, the front view, and the right side (end) view. The three basic views in the isometric drawing are like the three basic views in orthographic projection (see Chapter 16).

The first step in creating an isometric drawing is to establish the front right-hand corner of the

109

object. The drafter must remember to provide space to develop the front view to the left of this right-hand corner point, the right side (end of the object) from this corner point, and the plan view from this corner point.

To establish the right-hand corner, the drafter should visualize the drafting sheet divided into nine equal parts (three parts vertically and three parts horizontally (Figure 18.1). In the example, lines are numbered 1 through 4. Next, a construction line should be drawn along vertical Line 1. Line 1 will be the right-hand corner of the isometric drawing. The intersection of Line 1 and Line 3 is the front, bottom corner of the drawing. A point or dot should be placed at this intersection (Figure 18.2).

The drafter should draw a construction line from the point representing the bottom right-hand corner on the right-hand corner line. This construction line should intersect the point on the right-hand corner and project in an upward direction to the right at a 30-degree angle from the horizontal (Figure 18.3). This line will be the base of the front view.

The next step for the drafter is to draw a construction line to the left at a 60-degree angle from the horizontal. This construction line must intersect the right-hand corner line at the right-hand, bottom corner point and project up and to the left (at 60 degrees) from this right-hand, bottom corner point (see Figure 18.3). This 60-degree line will become the base for the front view (Figure 18.4).

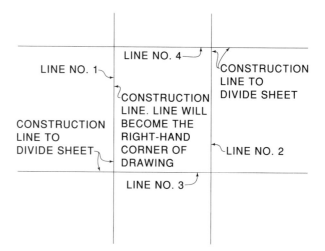

Figure 18.1   Dividing the space into three parts

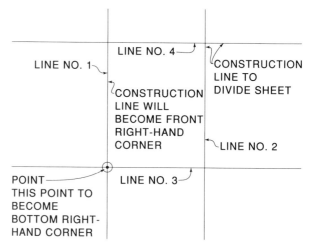

Figure 18.2   Locating the front, bottom corner of drawing

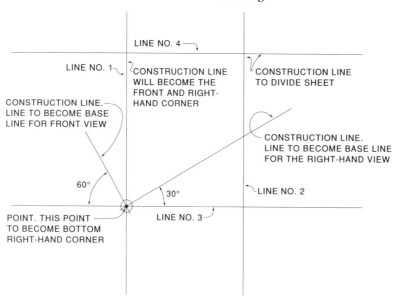

Figure 18.3   Establishing base lines for front view and right side view

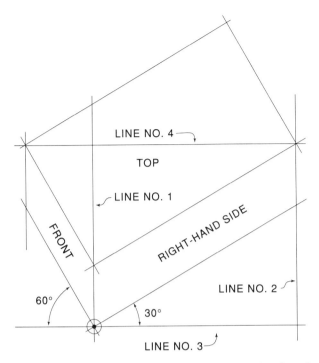

Figure 18.4 Establishing front corner and cube of the object

tances on the object where surfaces change directions and the object takes shape.

One important thing to remember is that *parallel lines are parallel to each other, and parallel surfaces are parallel to each other*. Figure 18.5 shows the object being developed. Known points are being identified and object lines are being drawn between these points. It is suggested that as a rule, one view of the object should be worked on at a time; however, in some cases, working on two views at the same time helps with the proportions of the object. Most drafters work to develop the plan view portion of the object first. This requires the development of the top profile of the rear view. When the front corner and the top view have been developed, the other two views are developed from these known items.

The finished drawing of the example is shown in Figure 18.6. This drawing was developed with 30°/60° = base angles.

With the front, right-hand corner established, and the base lines for the front view and the right side (end) view drawn, the drafter is ready to establish the height of the object, the width of the front view, and the length of the right side (end) view.

It should be noted at this point that the isometric drawing is not a drawn-to-scale drawing. It is, however, important to keep everything in proportion on the drawing. For this reason, the architect's scale is used to establish distances, but not for making the drawing *to scale*.

The drafter must use common sense to keep the drawing in proportion. For example, if the object being drawn is two times as long as it is wide, the drawing should reflect this difference. If the height of the object is half the width of the object, the drawing should reflect this difference. The architect's scale is a very useful tool to help establish these proportions.

After the length of the front view, the length of the right side view, and the height of the object have been established, the four corners of the isometric view can be established and a rectangular solid or a cube can be established with construction lines that set the outside limits of the object. Now, the drafter can use the architect's scale to establish dis-

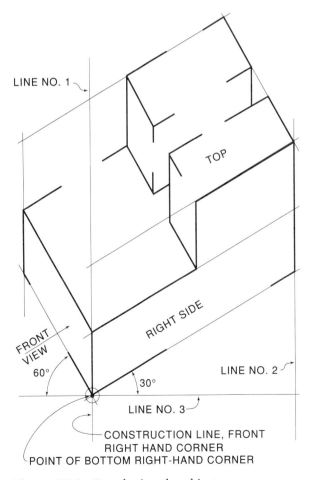

Figure 18.5 Developing the object

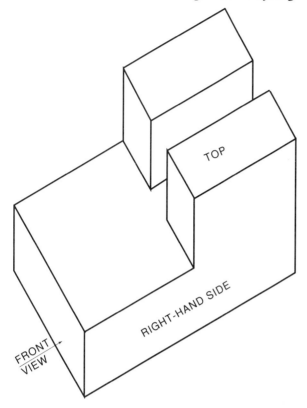

Figure 18.6 Finished drawing

18.3 A VARIATION OF ISOMETRIC DRAWINGS

In section 18.2 How Isometric Drawings are Drawn, the base lines were drawn from the right-hand corner (dot or point) at 30 degrees from the horizontal to the right, and 60 degrees from the horizontal to the left (Figure 18.3). These two construction lines become the base lines for the front and right side views of the objects.

Some isometric drawings can be developed with more detail if both of these construction lines are drawn at 30 degrees from the horizontal. In Figure 18.7, the construction lines are drawn from the right-hand corner point at 30 degrees from the horizontal. The procedure for dividing the drawing paper into six parts and the location of the right-hand corner point is the same as described for Figures 18.1 through 18.3 except for the angle of the base construction lines.

The construction lines establishing the outside limits of the object are shown in Figure 18.8. The rear outline of the object is developed, and the

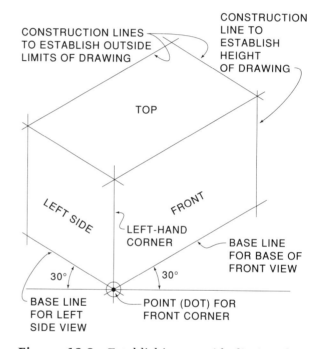

Figure 18.8 Establishing outside limits of an object

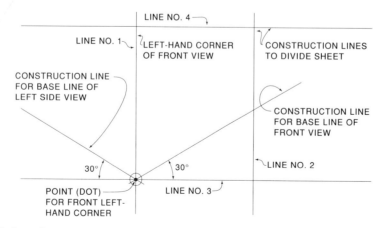

Figure 18.7 Establishing base lines for front and right-hand side views

plan view, front view, and the right side view are developed in the same manner as described at the beginning of this chapter (Figure 18.9). The finished drawing of the object is drawn with the base line for the front view and the right side view at 30 degrees from the horizontal (Figure 18.10).

When the two isometric drawings are compared (Figures 18.6 and 18.10), the right side view is about the same for each plan. The front view is larger in Figure 18.10 than in Figure 18.6 because the 30-degree base line makes the front view longer than in Figure 18.6, where the base line is shown at 60 degrees from the horizontal. The plan view (top) is larger in Figure 18.10 because of the 30 degrees from the horizontal for the base line in the front view.

The drafter must decide which angles to use when drawing an isometric plan. The best drawing is the drawing that will relay the most information to the worker or technician in the clearest fashion.

It may be necessary to rough out the base drawing up to the point of establishing the outside limits of the object (Figures 18.4 and 18.8) before making the decision as to which base angles should be used from the right-hand corner point.

18.4 ISOMETRIC PIPING DIAGRAM

An isometric drawing of a piping diagram is shown in Figure 18.11. This drawing shows a pump with the piping connections and the fittings, valves, and

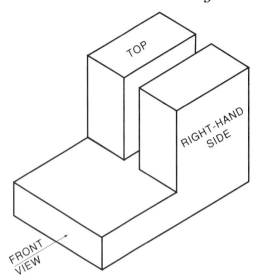

Figure 18.10   Finished drawing

equipment that should be installed in the piping system. Figure 18.11 was drawn using 30°/30° base lines from the horizontal.

Isometric drawings are used to great advantage in detailing piping diagrams. When isometric drawings are drawn in proportion, they show the relationships of the different components in the piping system, and they provide space to add notes, symbols, dimensions, and instructions as to how the system is to be installed.

18.5 WHERE ISOMETRIC DRAWINGS ARE USED

In the HVAC field, designers and drafters use isometric drawings for many different purposes. Some examples of the various ways that isometric drawings are used are described and illustrated on page 114. As mentioned earlier, Figure 18.11 is a piping diagram showing pipe fittings, special equipment, and where the equipment is to be installed. Figure 18.12 is a detail of a base for equipment. Figure 18.13 shows a fan mounted on an equipment base with ductwork connected to the fan. Figure 18.14 shows an offset in ductwork to avoid a pipe. Figure 18.15 shows piping connections to a heating coil located in an air-handling unit. Related controls and fittings are shown in the piping system. In many cases, the isometric drawing is the only place in a set of drawings where information is given to the installer or the craftsman detailing how the different components are to be installed.

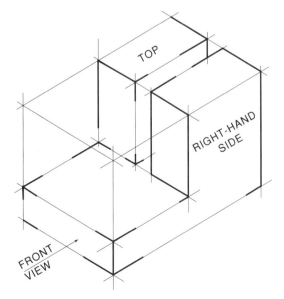

Figure 18.9   Developing the views of the object

AIR VENT

RELIEF VALVE

BOILER FITTING

EXPANSION TANK

AIR VENT

PUMP NO. 2
ZONE NO. 2

PUMP NO.3
ZONE NO. 3

BOILER

STUB-OUT
FOR FUTURE

PUMP NO.4
ZONE NO. 4

STUB-OUT
FOR FUTURE

BRACKET
BLOW-OFF LINE
TO BOILER JACKET

CONCRETE
HOUSEKEEPING PAD

GATE VALVE

UNION

BASE MTD. PUMP NO. 1
ZONE NO.1

CONCRETE PAD

Figure 18.11   Piping diagram

¹/₂ " ANCHOR BOLTS (THREADED)
ANCHORED IN CONCRETE 3"
WITH 1", 90° BEND IN BOLT
INSIDE CONCRETE PAD. BOLT
TO PROJECT 2¹/₂ " ABOVE PAD.
(FOUR REQUIRED)

3'-6"

1'-1"   1'-9"

6"

9"   3"

1'-8"

CONCRETE
HOUSEKEEPING
PAD

¹/₂" REINFORCEMENT BARS
6" ON CENTER EACH WAY

Figure 18.12   Detail of equipment base

SUPPLY DUCT

FLEXIBLE DUCT
CONNECTION-
4" CANVAS

"S" & DRIVE
CONNECTION

EXPANDED
METAL
INTAKE
GRILLE

12 X12   12 X12

RADIUS 90°
ELBOW WITH
TURNING VANES

CENTRIFUGAL
FAN UNIT

MOUNT ON 6" HIGH CONCRETE
HOUSEKEEPING PAD WITH ¹/₂"
REINFORCEMENT IN CENTER OF PAD.
REINFORCED 6" O.C. EACH WAY.
PROVIDE HOLDDOWN BOLTS TO FIT
FAN BASE.

Figure 18.13   Detail of ductwork and fan on
equipment base

HANG 14 X 6 DUCT AT THIS POINT
(TYPICAL FOR 4 HANGERS)

14 X 6 DUCT

6" ℄

14 X 6

14 X 6

OFFSET
14 X 6 DUCT UP 7"

6" ROUND DUCT

Figure 18.14   Detail of ductwork offset over pipe

SUPPLY MAIN
RETURN MAIN
GATE VALVE
AIR VENT
UNION
3-WAY CONTROL VALVE
BALANCING VALVE
UNIT
THERMOMETER
GLOBE VALVE
GATE VALVE
UNION
COIL IN UNIT

Figure 18.15   Piping connections to heating coil in air-handling unit

SUMMARY

• Although the oblique method of drawing is easier to *draw*, the isometric method of drawing is easier to *read*.

• Isometric drawings show the three basic views of an object (plan view, front view, and right-hand view).

• Although not drawn to scale, isometric drawings should be drawn in proportion.

• Parallel lines must be drawn parallel to each other, and parallel surfaces must be drawn parallel to each other.

• Often an isometric drawing is vital to show the craftsman how different components are to be installed.

• The drafter can choose whether construction lines are more effective when drawn at 30 degrees or 60 degrees from the horizontal.

REVIEW QUESTIONS

1. What are three methods of drawing pictorial representations of objects?

2. Which of the three methods listed above is the easiest to draw?

3. Which of the three methods listed above is the easiest to read?

4. What are the three basic views of an object that are shown in isometric drawings?

5. Is an isometric drawing done to scale?

6. What is the reason for using an architect's scale in creating an isometric drawing?

7. Fill in the blanks: parallel lines should be drawn _____ ____ _____ _____, and parallel surfaces should be drawn _____ ____ _____ _____.

8. Give at least two reasons why isometric drawings can be advantageously used in detailing piping diagrams.

ADDITIONAL STUDENT EXERCISES: Freehand Sketching
Isometric Views

1. Using graph paper, freehand sketch isometric views of the objects shown in Figure 18.16, using orthographic projection.

2. Draw isometric views of the orthographic objects shown in Figure 18.17. Use orthographic projection.

3. Freehand draw an isometric drawing of the air-cooled condensing unit shown in Figure 18.18, using orthographic projection. The unit is mounted on a concrete pad and has a round, vertical air discharge *with* a round fan guard.

4. Freehand draw an isometric drawing of the gas-fired furnace shown in Figure 18.19, using orthographic projection.

5. Refer to Figure 18.11 to answer the following questions.
   A. Are unions provided for quick pump disconnect?
   B. How many zones are being served by the boiler?
   C. Is a base-mounted pump shown?
   D. Does the boiler have a housekeeping pad?
   E. Have provisions been made for future zones?
   F. How many safety valves are shown?
   G. Is the piping shown located underfloor?
   H. Have provisions been made to remove air from the system?
   I. How many gate valves are shown on the drawing?
   J. Is an expansion tank shown?

6. Refer to Figure 18.12 to answer the following questions.
   A. What is the over-all dimension of the equipment base?
   B. What is the size of the reinforcement bars in the pad?
   C. Are there two rows of reinforcement bars?
   D. How thick is the pad to be?
   E. What is the size of the anchor bolts?
   F. How far are the bolts located from the edge of the pad? (two answers)

7. Refer to Figure 18.13 to answer the following questions.
   A. What kind of connection is used in the ductwork?
   B. Does the fan have vibration isolation on the duct connection?
   C. Does the fan have a housekeeping pad?
   D. Is the discharge ductwork insulated?
   E. What is the thickness of the housekeeping pad?
   F. Does the concrete pad have reinforcement?
   G. What is the size of the reinforcement and how is it spaced?
   H. What is the intake opening to the fan covered with?
   I. Does the fan have hold-down bolts?

8. Refer to Figure 18.14 to answer the following questions.

   A. Why is the offset needed?

   B. Does the ductwork have hangers?

   C. Are the types of duct joints shown?

9. Refer to Figure 18.15 to answer the following questions.

   A. Is the heating coil indicated?

   B. How many gate valves are shown?

   C. How many thermometers are shown?

   D. Is there a globe valve in the system?

   E. Is there a balancing valve shown?

   F. How many unions are shown?

   G. Is the direction of flow indicated?

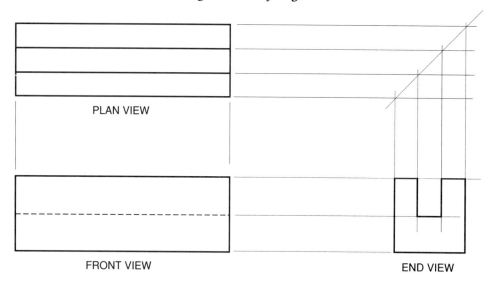

PLAN VIEW

FRONT VIEW

END VIEW

DIRECTIONS: DRAW ISOMETRIC VIEW OF ORTHOGRAPHIC FIGURE ABOVE.

Figure 18.16 Exercise 1

PLAN VIEW

FRONT VIEW

END VIEW

DIRECTIONS: DRAW ISOMETRIC VIEW OF ORTHOGRAPHIC OBJECT ABOVE.

Figure 18.17 Exercise 2

DIRECTIONS: FREEHAND DRAW AN ISOMETRIC DRAWING OF THE AIR-COOLED CONDENSING UNIT SHOWN ABOVE. THE UNIT IS MOUNTED ON A CONCRETE PAD, AND HAS A ROUND, VERTICAL AIR DISCHARGE WITH A ROUND FAN GUARD.

Figure 18.18   Exercise 3

PLENUM

GAS VENT

FLEXIBLE DUCT CONNECTION

FURNACE

CONCRETE PAD

DIRECTIONS: FREEHAND DRAW AN ISOMETRIC DRAWING OF THE GAS-FIRED FURNACE
USING ORTHOGRAPHIC PROJECTION.

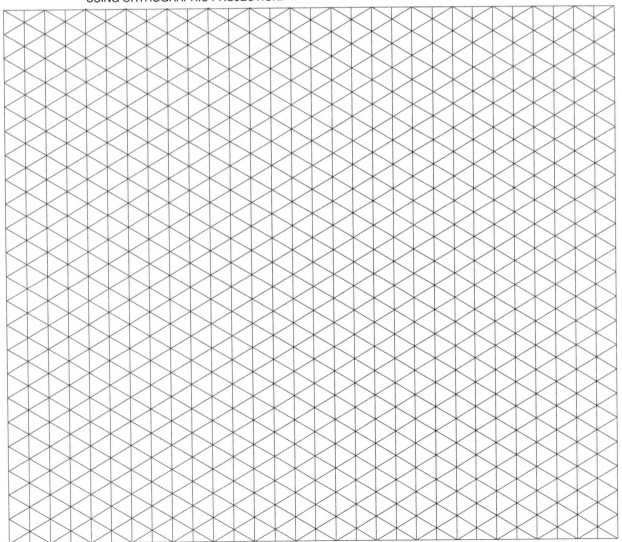

Figure 18.19 Exercise 4

LETTERING AND NOTATIONS

OBJECTIVES

After studying this chapter, you should be able to:

■ discuss the use of upper- and lowercase letters

■ demonstrate the ability to draw vertical-style letters and numerals

■ demonstrate the ability to draw sloped-style letters and numerals

■ describe and illustrate the proper spacing between letters, words, and sentences

■ discuss factors that are considered in locating notes on plans

■ define a leader and demonstrate its use on drawings

19.1 INTRODUCTION

It is sometimes said that the lettering on a drawing makes or breaks the whole drawing. While accuracy and clarity of the drawing itself are very important, the lettering and notations affect the overall appearance of the drawing and relay much necessary information. The beginning drafter soon learns that a beautiful drawing can become a mediocre or poor drawing if the lettering of notes and titles is of poor quality.

This chapter will discuss several types of lettering commonly used on drawings and will give examples of vertical and sloped lettering. Achieving high quality lettering in manual drafting requires repeated and consistent practice. When computer drafting is being used, the lettering can be selected and formed on the plan in an almost perfect manner.

Proper notation is also covered in this chapter. The wording and location of notes will be discussed along with how the notes should relate to the drawing.

19.2 UPPERCASE LETTERING

The two basic types of letters used on working drawings are uppercase letters and lowercase letters. Uppercase letters are used most often. These letters can be formed in vertical (Figure 19.1) or sloped style (Figure 19.2). Engineers often use sloped letters, and architects seem to prefer vertical letters.

In Figures 19.1 and 19.2, the small arrows indicate how letters are to be formed. For example, to form the "F," the first stroke of the pen or pencil should be the vertical portion of the "F," followed by the second stroke to form the top of the letter, with the last stroke forming the middle line. The drafter must visualize circles while forming the curved portions of the letters O, C, G, D, B, R, J, U, and S. Practice is the secret to good lettering.

Sloped letters are formed in the same manner as vertical letters except that they are sloped some 20 degrees off vertical (Figure 19.2).

Many architects and engineers follow the basic system for forming letters; however, many add their own frill to the letters to give them a special appearance. This practice is acceptable if the owner approves.

Figure 19.1 Vertical uppercase letters

19.3 LOWERCASE LETTERING

The suggested method for forming lowercase letters is shown in Figure 19.3. The small arrows on Figure 19.3 indicate the suggested strokes for forming the letters. More circular strokes are required for forming lowercase letters than for uppercase. These circular strokes require practice to improve the quality of the letters. Lowercase letters can be vertical or sloped style (see Figure 19.3).

Some owners prefer a combination of uppercase and lowercase letters. The federal government usually requires combination-style lettering. The combination style has the first letter of a sentence in uppercase and the remaining letters in the sentence in lowercase. If sloped lettering is being used, lowercase letters are sloped the same as uppercase letters.

19.4 LETTERING NUMERALS

Numerals (symbols for numbers) should be formed as shown in the same style as letters. If letters are vertical, numerals should be vertical. If letters are sloped, numerals should be sloped. The suggested method for forming slant numerals is illustrated in Figure 19.4, page 126. Numerals also may require several partial circles for proper formation. Here again, practice is required to form numerals well.

Fractions may become a problem when they are combined within a line of lettering. Fractions can be formed two ways (Figure 19.5, page 126, for examples). The first method is to establish the dividing line in the center of the lettering line and place the fractional letters above and below this line. Figure 19.5A demonstrates this method. The

Figure 19.2   Sloped uppercase letters

VERTICAL LETTERS

SLOPED LETTERS

Figure 19.3   Lowercase letters

Figure 19.4 Forming slant numerals

numerals should be the same size as the letters between the lettering guidelines. Another way is to write the numerator of the fraction between the guidelines, place a sloped line after the numerator, and place the denominator after the sloped line (Figure 19.5B). The two numerals should be drawn the same size as the lettering. Figures 19.5A and 19.5B show sloped numerals. Vertical numerals are formed in the same manner.

19.5 SIZE AND SPACING OF LETTERS AND NUMERALS

Lightly drawn guidelines regulate the height of each letter or numeral. The use of a lettering guide is suggested. Several devices are available; however, the adjustable lettering guide is more commonly used (Figure 19.6). Figure 19.6 shows how guidelines are drawn using the lettering guide. Figure 19.7 shows how the lettering guide is used to establish guidelines for sloped letters (20 degrees off center). Figure 19.8 shows how the lettering guide can be used to establish guidelines for vertical lettering.

The circular portion of the lettering guide can be rotated to space the lettering guideline closer or further apart. The series of holes provides the drafter the capability of choosing the spacing pattern for the guidelines.

Figure 19.5 Lettering fractions

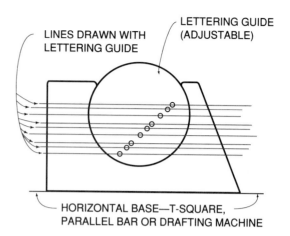

Figure 19.6 Lettering guidelines drawn with a lettering guide

Spacing between the letters in a word should measure approximately one-fourth the width of a letter. Spacing between words should be about two-thirds the width of a letter. Spacing between sentences should be approximately one and one-half times the width of a letter. These distances are suggested to serve as a guide to the drafter, who can establish a method of spacing letters and numbers to keep notes and sentences uniform.

The beginner drafter will find that a pencil with soft lead is easier to use for lettering a drawing. The point should be moderately sharp. Experimenting with different pencils and practicing often enables each drafter to find the pencil best suited to fit his or her needs.

Remember that some architects and engineers require that certain styles of lettering be used in

Figure 19.7. Lettering guide used to draw guidelines for sloped letters

GUIDELINES FOR VERTICAL LETTERS

LETTERING GUIDE (ADJUSTABLE)

LETTERING LINES

HORIZONTAL BASE—T-SQUARE, PARALLEL BAR OR DRAFTING MACHINE

Figure 19.8 Lettering guide used to draw guidelines for vertical letters

their offices. The beginning drafter needs to learn the standard style of lettering within that office. Knowing how to form the letters and numerals using the vertical and slope methods helps the beginner to learn any special lettering styles.

19.6 WORDING NOTATIONS FOR BLUEPRINTS

The architect, engineer, and drafter are responsible for relaying design information to the workman or installer. Plans show where the work is to be performed and what materials are to be used, but additional instructions are often needed. This information is covered by notes. To save space and keep the plan from becoming cluttered, notes must be written in precise language.

It is also important to locate the notation in a logical place on the drawing. The note should be close to the area where it applies. Notes are written instructions that tell the types of materials, sequence of installation, how different members fit together, equipment usage, how to operate the equipment, and other pertinent information.

To link the note to the object it is referred to, a thin line called a leader is extended from the beginning or the end of the notation to the place where the note applies. The leader usually starts near the notation and extends to the reference object, terminating with an arrowhead. The leader is usually drawn straight out from the note and then turned at an angle to the reference object and arrowhead. Notes should always be located so that the leader lines do not cross. A typical note with leader is shown in Figure 19.9. Some architects and engineers draw a freehand leader. Freehand leaders are used in the set of plans at the end of this textbook.

In some cases, notes are too long to place in the location where they are needed. In that case, the drafter will insert brief instructions such as "See Note No. 1." These long notes are then lettered on a blank space on the plan so that the floor plan is not cluttered.

Most notes are lettered horizontally, but sometimes it is advantageous to letter the note vertically. When vertical notes are used, they should be readable from the right-hand side of the plan (Figure 19.10).

THIS IS AN EXAMPLE OF A NOTE AND A LEADER

Figure 19.9 Typical note with leader

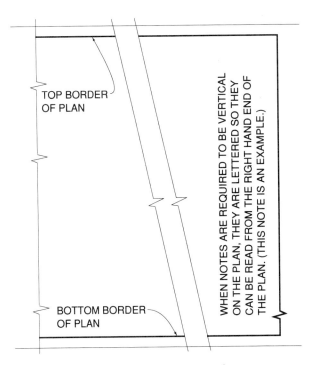

TOP BORDER OF PLAN

BOTTOM BORDER OF PLAN

WHEN NOTES ARE REQUIRED TO BE VERTICAL ON THE PLAN, THEY ARE LETTERED SO THEY CAN BE READ FROM THE RIGHT HAND END OF THE PLAN. (THIS NOTE IS AN EXAMPLE.)

Figure 19.10 A vertical note on a plan

SUMMARY

• Two basic types of letterings are all uppercase letters and combination upper- and lowercase letters.

• Vertical letters are formed from top to bottom, and sloped letters are tilted about 20 degrees off vertical.

• The same style should be used for letters and numerals, if appearig in the same drawings.

• The drafter should use guidelines to regulate the height of letters and should practice spacing lettering in a neat and attractive way.

• If possible, notes should be located close to the area to which they refer with a leader connecting the note to the proper area.

REVIEW QUESTIONS

1. Following the guidelines in this chapter, print the letters of the alphabet in uppercase letters, vertical style.

2. Following the guidelines in this chapter, print the letters of the alphabet in lowercase letters, vertical style.

3. Following the guidelines in this chapter, print the letters of the alphabet in uppercase letters, sloped style.

4. Following the guidelines in this chapter, print the letters of the alphabet in lowercase letters, sloped style.

5. Print the numerals 1 through 20 in vertical and sloped style.

6. Define a leader.

7. How should a drafter use guidelines?

8. What is the purpose of a lettering guide?

9. Print a paragraph from this chapter demonstrating acceptable spacing between letters, words, and sentences.

10. Print, in an acceptable style, three rules for proper notation on plans.

ADDITIONAL STUDENT EXERCISES: Practice Forming Letters

1. Using the grids provided in Figure 19.11, practice forming vertical letters.

2. Practice forming sloped letters using the grids provided in Figure 19.12.

VERTICAL LETTERS

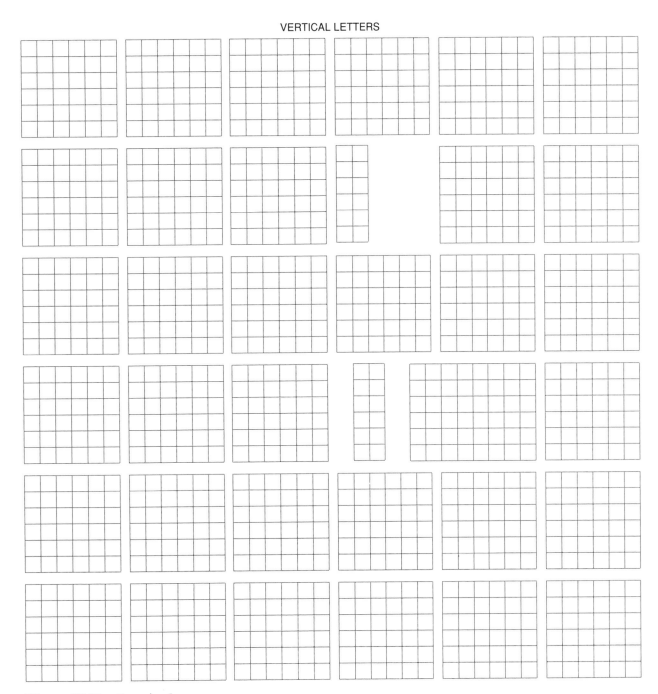

Figure 19.11   Exercise 1

SLOPED LETTERS

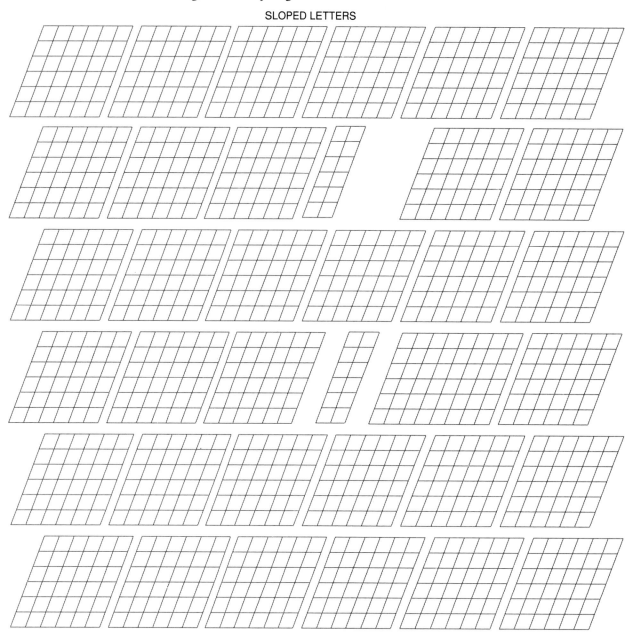

Figure 19.12   Exercise 2

ORGANIZING A DRAWING SHEET

OBJECTIVES

After studying this chapter, you should be able to:

- describe how a basic schedule of plans is set up
- discuss the information put on planning sheets
- explain the purpose of a completion log
- tell how a drafter determines the placement of a drawing or drawings on a sheet
- tell why it is important to draw neat, uncluttered plans

20.1 INTRODUCTION

The designer and drafter are responsible for relaying information to the construction worker or technician in a thorough, neat, and orderly manner. To accomplish this task, they must first plan what is to be shown on the sheet and then decide how to arrange this information in a logical manner for the worker or technician. This chapter will give suggestions for the organization of a drawing sheet.

Architectural and engineering offices usually have their own procedures for planning each sheet of drawings that they produce. Some of these procedures will be discussed in this chapter.

20.2 CREATING A BASIC SCHEDULE OF PLANS

At the beginning of a job, the architect and engineer usually call a meeting of the designers and drafters who will be working on the project. The purpose of this meeting is to plan the drawings needed to relay the necessary information to the workmen or technicians. The number of sheets and the titles of these sheets are listed. The general grouping and numbering of the plans are considered. For example, the site plan or plans come first, followed by the floor plans, elevations, sections, and details. The schedules are the last group of plans.

The size of the job determines the number of sheets of drawings that will be required. The larger the project, the more planning is required to insure that all essential information is included on the drawings. Larger jobs also require more coordination among different sheets of drawings.

20.3 A SYSTEM FOR ORGANIZING A JOB

A popular system for organizing plans for a project uses an 8½-by-11-inch sheet of paper to represent each sheet of drawing that is required. The sheet number is placed at the lower right-hand corner of each 8½-by-11-inch sheet. The sheet number corresponds to the sheet number in the regular drawings. All drawings, details, elevations, sections, and schedules are listed on these sheets. At the end of the planning sessions, the architect or engineer knows *approximately* how many sheets are required to design and draw the plans for the project, but the *exact* number of sheets will not be known until they are actually drawn. During the design, additional sheets may have to be added to relay all of the required information.

If a sheet is added to the drawings, an 8$\frac{1}{2}$-by-11-inch sheet should be added to the original planning booklet. Sheet numbers (and sometimes sheet titles) should be changed to reflect the addition of the sheet. The planning booklet containing the 8$\frac{1}{2}$-by-11-inch planning sheets should always be kept up-to-date. This way the design team will always know what sheets are required as well as what information is required on each one.

In larger offices, each designer may be assigned a certain sheet to complete. With a project planning booklet, each designer knows what information is needed on that particular sheet and can assemble the information to make the design.

Coordination with other designers and drafters should take place on a regular basis to insure that the overall systems will work, that the design work does not overlap, and that important pieces of design work are not being omitted.

A completion log is usually kept on the 8$\frac{1}{2}$-by-11-inch sheets. When this log is kept up-to-date, each member of the team can look at the planning book to see what sheets have been finished or what percentage of each sheet has been finished. This log helps keep the project on schedule because additional manpower can be added if the job is lagging behind schedule.

20.4 ORGANIZING INDIVIDUAL DRAWING SHEETS

Each drawing sheet must have a title for identification. Once this title has been established, all the information shown on the sheet should pertain to this title. For example, if the sheet title is "First Floor Plan," the first floor plan with all the information related to the first floor plan should be shown on the sheet. Often the floor plan does not fill the full sheet. It is permissible to draw details, sections, and elevations pertaining to the floor plan on the floor plan sheet. In this case, the sheet title should be changed to include the details, sections, and elevations.

In the example in the preceding paragraph, the floor plan is the most important drawing on the sheet; therefore, it should be shown in a prominent place, usually the center of the sheet. Some architects and engineers prefer to center the floor plan on the sheet from top to bottom but shifted to the right-hand border rather than the center of the sheet from side to side. This placement allows more usable space to the left of the floor plan for details, sections, and elevations.

The most important thing to remember in planning a drawing sheet is that the sheet must be organized and easy to read. Cluttered and unorganized sheets are hard to read and may cause costly mistakes.

SUMMARY

• The architect, engineer, designers, and drafters on a job hold a planning meeting before any drawing is done.

• The larger the job, the more sheets of plans and the more coordination is needed.

• Many firms make a planning booklet with one 8$\frac{1}{2}$-by-11-inch sheet corresponding to each proposed sheet of drawings.

• The planning booklet must be kept up-to-date regarding the number of sheets of drawings and what information is to be put on each sheet.

• A completion log keeps workers informed as to what percentage of the drawings on each sheet has been finished at any given time.

• The most important drawing on a sheet should be the most prominent item on the page.

• If more than one drawing is put on a sheet, the title for that sheet should reflect the contents of the sheet.

• Each sheet should be neat, uncluttered, well-organized, and easy to read.

REVIEW QUESTIONS

1. What takes place at the planning meeting before any drawing is done?

2. Who attends this planning meeting?

3. If a planning booklet is prepared as discussed in this chapter, what information goes on each 8½-by-11-inch sheet?

4. What is the purpose of a completion log?

5. How can a person know whether more than one type of drawing has been put on the same sheet?

6. Why is it important to have each sheet uncluttered, neat, and well-organized?

ADDITIONAL STUDENT EXERCISES: Constructing Project Plan Booklets

1. Review the construction plans for "A Small Bank and Trust Company" included at the end of the book. Construct a project plan booklet for the design of the building. Create a sheet for each sheet of drawings, and describe the drawings that have been drawn on each sheet. Add or omit any items that may improve the quality of the set of plans. Details and sections may be moved from one sheet to another. Give at least one reason for making such changes.

2. After reviewing Chapter 11, "Construction Drawings," and reviewing the plans at the end of this textbook, you should be familiar with the various sections of a set of plans and what is generally required for each section.

   Visualize a set of plans for a residence of approximately 2,000 square feet with three bedrooms, two bathrooms, a family room, kitchen, dining room, and living room.

   Construct a project plan booklet for this residence.

21

SHADING AND CROSSHATCHING

OBJECTIVES

After studying this chapter, you should be able to:

- ■ recognize and define crosshatching
- ■ recognize and define shading
- ■ discuss when and why shading, crosshatching, and poché are used on drawings
- ■ explain how shading is used to heighten the three-dimensional effect on drawings
- ■ discuss the use of commercially prepared stick-on shading to achieve a uniform appearance on drawings

21.1 INTRODUCTION

When a drawing has been completed, it is sometimes hard to distinguish the object line separating two objects or two different materials. To help with this problem, crosshatching or shading can be applied to the object to help identify the dividing line. Poché is a type of shading that makes the dividing line between objects more apparent. This procedure is used often throughout a set of building plans to identify different materials used in construction. Shading makes the three-dimensional effect much more pronounced in oblique and isometric drawings. In this chapter, the topics of crosshatching and shading will be discussed and demonstrated.

21.2 CROSSHATCHING

Crosshatching is accomplished by using a 45-degree triangle or a 30/60-degree triangle to draw parallel lines, equally spaced, over a selected area (Figure 21.1). A 45-degree angle is commonly selected for manual crosshatching. A drafting machine can also be used for crosshatching. The drafting machine must be set at the desired angle and parallel lines, equally spaced, drawn over a selected area. The crosshatching lines are drawn from one object line to the other object line.

The quality of the lines should be the same as the extension and dimension lines. These lines are usually thin lines; however, they must be heavy enough to print (Chapter 15, "Types and Weights

Figure 21.1 Crosshatching at different angles

of Lines"). In some cases, crosshatching lines are drawn the same weight as object lines. The choice of line weights for crosshatching is usually left up to the drafter.

For crosshatching, the parallel lines can be drawn in any direction and at any angle. If two kinds of materials (or two different pieces) fit together, crosshatching can be used to help separate them (Figure 21.2). The drafter should turn the crosshatching lines in one direction from the common object line and then turn the crosshatching lines in another direction from the same object line (Figure 21.3).

21.3 POCHÉ

Poché (pronounced pō **shay**) can be defined as the process of shading an object with a soft graphite pencil and rubbing the graphite with a finger or a soft tissue to create a uniform shading effect. This procedure has been used for many years to distinguish between different materials on building drawings. One example of the use of poché is the treatment of partition walls on an architectural floor plan (Figure 21.4). When the walls have poché, they are easily seen because of the contrast between the walls and the floor.

Figure 21.2 Two different metals indicated by crosshatching

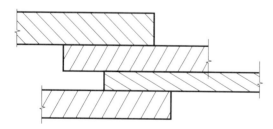

Figure 21.3 Crosshatching of different materials

Figure 21.4 Partition wall with poché

When soft graphite is rubbed, it is sometimes spread outside the object lines. If this happens, the drafter must clean the area outside the poché with an eraser.

Poché can be used in conjunction with other drafting methods. When the architect draws concrete block walls, the symbol for concrete block is put on the plans. Poché is added to the symbol for concrete block because it makes the walls look more realistic (Figure 21.5). Concrete walls are often shown with the concrete symbol, and poché is added to give the walls more contrast.

21.4 COMBINATION OF CROSS-HATCHING AND POCHÉ

It is often advantageous to use several different methods to distinguish between different materials or parts that are separated by only one object line. To make this distinction clear, crosshatching can be used on one material or part. Crosshatching can also be used for several different materials so long as the parallel lines are drawn at different angles. To break the monotony, some kind of poché can be used in conjunction with the crosshatching, or poché can be used between the sections of crosshatching (Figure 21.6).

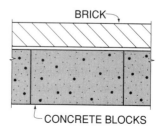

Figure 21.5 Brick and concrete block wall with concrete blocks shown with poché

Figure 21.6 Pipe flanges with gasket (gasket shown poché)

21.5 SHADING DRAWINGS

In Chapters 17 and 18, oblique and isometric drawings are described. Oblique and isometric drawings are very important in the air conditioning and heating fields, because they are pictorial representations that give a three-dimensional effect. With shading, the three-dimensional effect is greatly improved.

21.6 PROCEDURES FOR SHADING OBLIQUE AND ISOMETRIC DRAWINGS

A drafter who has finished an oblique or isometric drawing can improve the drawing by shading the object in the drawing. First, the drafter must imagine a floodlight being played on the object (Figure 21.7). When this light strikes a surface of the object, the surface becomes brighter and shadows are cast on surfaces that are not struck by the light. These shadows appear darker than the lighted surfaces.

Keeping the imaginary light in mind, the drafter should mark the shaded areas on the drawing. The shaded areas are the areas where shadows would result if a light were shown on the object. These markings should be very light so that they can be erased. This imaginary lighting should always strike the object at an angle (Figure 21.8). A 45-degree angle is commonly used; however, other angles are

Figure 21.7 Shading resulting from light placed on object

sometimes used with good results. The drafter may try several angles for the imaginary light.

When all shaded areas on the object have been marked, the drafter decides what symbol (crosshatching or poché) is to be used to indicate the shaded area. Special care should always be taken to make the shading uniform.

Commercially prepared, stick-on shading can be used for the shading. This stick-on shading is a transparent sheet with printed dots on one side. The other side of the transparent sheet has a colorless glue that adheres to the drafting paper. When the sheet is cut to fit the shaded area, the backing is peeled off and the prepared shading is applied as needed on the drawing. When the drawing is run through the printing machine, the dots of the stick-on sheet show up the same as if they were drawn by hand. This procedure gives the finished drawing a uniform shading or shadowing that is neat and easily read.

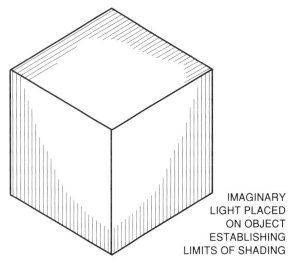

Figure 21.8 Marking limits for shading

SUMMARY

• Crosshatching, shading, and poché are used to make the dividing line between objects on a drawing more apparent.

• Crosshatching lines are thin lines drawn close together at an angle (typically a 45-degree angle) from object line to object line.

• Poché is the procedure by which an object is shaded with a soft graphite pencil and then rubbed with a tissue or the finger to achieve uniform shading.

• Shading improves the three-dimensional effect on isometric and oblique drawings.

• To shade accurately, the drafter must duplicate the effect created by a floodlight shining on one surface of an object and casting shadows on other surfaces.

• A uniform shading effect can be achieved by the use of commercially prepared stick-on paper.

REVIEW QUESTIONS

1. When should a drafter use crosshatching and shading?

2. Why is shading important on oblique and isometric drawings?

3. How should the triangle be used to make crosshatching?

4. How can a drafting machine be used for crosshatching?

5. What criteria are used for determining the weight of crosshatching lines?

6. Why are crosshatching and poché sometimes used together?

7. Define poché, and give an example of when it is used.

8. Describe the procedure by which a floodlight defines areas to be shaded.

9. What is the advantage of using commercially prepared stick-on paper for shading?

ADDITIONAL STUDENT EXERCISES: Shading and Crosshatching

1. Using the examples illustrated in Figure 21.7 and 21.8, shade the vertical tank in Figure 21.9.

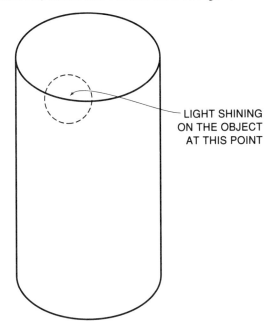

Figure 21.9   Exercise 1

2. Using the examples illustrated in Figure 21.7 and 21.8, shade the unit depicted in Figure 21.10.

Figure 21.10   Exercise 2

3. Using the examples illustrated in Figure 21.7 and 21.8, shade the horizontal tank shown in Figure 21.11.

Figure 21.11   Exercise 3

4. Referring to Figure 21.12, choose the light source spot where light strikes the gas furnace, and crosshatch to indicate shading.

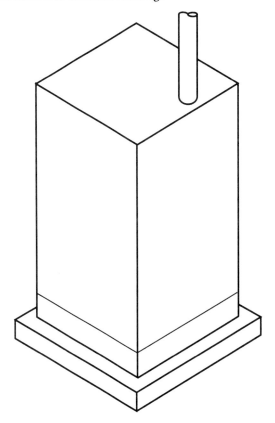

Figure 21.12   Exercise 4

5. Determine the light source spot where the light strikes and shade the object shown in Figure 21.13 using the poché method.

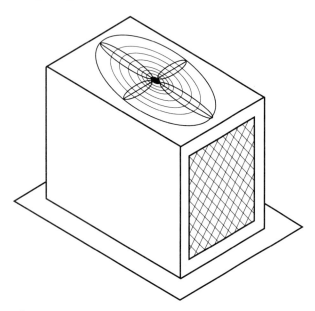

Figure 21.13 Exercise 5

22

DRAWING CONSTRUCTION ELEVATIONS

OBJECTIVES

After studying this chapter, you should be able to:

- describe typical information found on exterior elevations
- locate and discuss cutting plane lines and symbols used to denote cutting planes
- identify interior wall elevations
- describe typical information found on interior wall elevations
- describe information found on mechanical, electrical, and plumbing elevations

22.1 INTRODUCTION

Many types of elevations are required in building construction. Outside elevations of a building include front, rear, and side elevations. Interior wall elevations include kitchen cabinet elevations and built-in specialties. The mechanical, plumbing, and electrical sections in a set of building plans include elevations of equipment, equipment rooms, and other features. The larger the job, the more elevations are required to show how the finished product will look, the location of the equipment, and the specialty items required with the equipment.

22.2 EXTERIOR ELEVATIONS

Exterior elevations are orthographic views of the various sides of a building. Often the architect identifies the view by the direction it faces. Thus if the front elevation faces north, it may be called the north elevation, and the rear view would be called the south elevation. Other architects simply identify the front, rear, and side views of the drawings without including the direction each one faces as part of the title.

The architect or drafter shows the elevations of the various sides of the building in the architectural section of the plans. These drawings are developed from the floor plans of the building (Figure 22.1). The doors, windows, and so forth are located at

Figure 22.1 Developing an elevation from a floor plan

the same place on the elevation as on the floor plan. Types of windows and doors are drawn in detail so that the owner and the builder will know what they will look like. The windows and doors are also described in detail in the schedule section of the plans. Dimensions for windows and doors are usually given in these schedules.

On Sheet A-8 of the plans at the end of the book is an example of an elevation developed from a partial floor plan (see "Plan of 'B' Wall"). The elevation of the wall, developed from the partial floor plan, is shown in the lower left corner of the sheet.

Sheet A-2 of this set shows the exterior elevations of a small bank and trust company. The front elevation is not shown on Sheet A-2, as such, but the note under the rear elevation states that the front elevation is the same as the rear elevation except that it is reversed.

The cutting planes for sections drawn at this point are represented by circles with two numbers located inside, with a long line projecting through the height of the building. The top number in the circle represents the section number and the bottom number gives the page number where the section is drawn.

Sheet A-5 of the drawings shows transverse and longitudinal sections of a bank building. These sectional drawings supplement the information on the exterior and interior building elevations with detailed information on the doors, glass, and walls. The windows are fixed glass, requiring no schedule, and the doors are shown on the floor plan (Sheet A-1). The doors are numbered and are scheduled on Sheet A-8. If windows other than fixed glass had been used, the windows would have been numbered on the floor plan (and sometimes on the elevation plan). The windows would have been scheduled on the same sheet as the door schedule.

Other items that are usually shown on the outside elevation plans include hose bibbs, downspouts, openings in the outside wall, and all pieces of equipment mounted on the wall. Any features that can be seen on the finished outside wall should be shown on the outside elevation drawing.

22.3 INTERIOR ELEVATIONS

Interior elevations are very important to the owner and the builder. Interior elevations are usually drawn for cabinets and special built-in fixtures. They normally show everything that will be located on the finished wall. The set of construction drawings at the back of the book includes interior elevations of the teller's counter. These elevations, shown on Sheet A-7, are drawn in conjunction with the floor plan so that all dimensions for constructing the cabinets are shown on one of the two drawings. Another wall elevation is shown on Sheet A-8.

Interior elevations of this type are usually drawn for each wall in the building that requires special construction or special features on that wall. Interior wall elevations usually are drawn from one adjoining wall to the other adjoining wall. Items shown include electrical outlets, chair rails, wall finishes, shelving, crown molding, baseboards, telephone outlets, and special wall-mounted devices such as safes, fire extinguishers, and special wall features. An interior elevation of a wall of a bathroom might show built-in cabinets, mirror, lighting, toilet fixtures, electrical receptacles, and shelving. Specific details are shown on interior elevations to help the builder and tradespeople know what the finished product should look like. On larger jobs, several sheets of drawings may be required to accommodate all the required wall sections.

22.4 MECHANICAL ELEVATIONS

Mechanical elevations, like architectural elevations, show how equipment should look when the installation is completed. The engineer or drafter draws the equipment and shows all connections to the equipment. Elevations in the mechanical section include elevations of cooling towers, air-handling equipment, compressors, chillers, special ductwork, ductwork connections, piping, condensing units, and any other mechanical equipment requiring detailed instruction for installation.

Interior wall elevations are sometimes required to show supply and return air outlet installations and how they fit into the room decor. An equipment elevation of a condensing unit is shown on Sheet M-1 of the construction plans at the end of the book. Drawing 5 on Sheet M-1 shows the condensing unit and the mounting arrangement. The larger the job, the more mechanical elevations are required to show the required information.

22.5   ELECTRICAL ELEVATIONS

The electrical engineer and drafter use elevation drawings to show the location and to describe the mounting of electrical equipment. It may be advantageous for the engineer or drafter to draw interior wall elevations of an electrical room to show where each piece of equipment is to be located (Figure 22.2).

A wall elevation inside the building can show the location of electrical outlets, telephone connection jacks, and wall-mounted lighting fixtures.

An exterior elevation drawing may be required to show the electrical service wiring, the weatherhead, and the electrical meter location (see Figure 22.3). This drawing can also show how the service is to be grounded.

22.6   PLUMBING ELEVATIONS

Plumbing elevations are as important as any other elevations; however, not as many elevations are usually required in the plumbing trade. On larger jobs, the engineer or drafter sometimes draws elevations of toilet spaces and equipment room walls to show location of plumbing equipment.

Figure 22.2   Elevation of north wall of electrical room

Figure 22.3   The electrical weatherhead provides a waterproof service entrance for the electrical wires to the structure

SUMMARY

• The larger the job, the more elevations are required to show what the finished product should look like.

• Exterior elevations can be identified by the view (front, rear, right side, or left side) or by the direction faced (north, south, east, or west).

• Elevation views are developed from the floor plans of the building.

• Cutting planes are used to cut away sections of the building and show details of construction.

• The exterior elevation should show everything that can be seen on the finished outside wall.

• Interior elevations are usually drawn from one wall to the other to show specific details of construction on that wall.

• Mechanical and electrical elevations show how equipment should appear when installation is completed.

• Plumbing elevations are drawn to show locations and details of plumbing equipment.

REVIEW QUESTIONS

1. An exterior drawing of the front of a building facing south can be titled _____ _____ or _____ _____ .

2. A detailed description of windows and doors is given in the _____ section of the plans.

3. Circles denoting the location of cutting planes contain two numbers; the top number represents the _____ _____ and the bottom number represents the _____ _____ .

4. Hose bibbs and downspouts are shown on the (exterior, interior) elevation drawings.

5. Cabinets and built-in features are shown on the (exterior, interior) elevations.

6. Interior elevations are usually drawn from (wall to wall, floor to ceiling, outside to inside).

7. Electrical outlets, safes, and chair rails are shown on (exterior, interior) elevation drawings.

8. Special ductwork is drawn on the (mechanical, electrical, plumbing) elevations.

9. Wall-mounted lighting fixtures are shown on the (mechanical, electrical, plumbing) elevations.

10. Toilet equipment is shown on the (mechanical, electrical, plumbing) elevation drawings.

23

DRAWING CONSTRUCTION DETAILS

OBJECTIVES

After studying this chapter, you should be able to:

- explain why details are needed for coordination on the job
- identify commonly used architectural details, and discuss information on them
- identify commonly used mechanical details, and discuss information on them
- tell how to locate details on drawings
- describe the process by which details are usually developed

23.1 INTRODUCTION

The construction and equipment for the modern buildings of today have become so complex that no one person can plan the entire building. The architect, engineer, and designers/drafters are dependent on each other for answers to questions as to strength, space, construction, mechanical apparatus, design, and appearance of the finished product. The architect is usually associated with architectural engineers (structural, mechanical, plumbing, and electrical) who work closely to coordinate the design with the various trades through joint meetings and open discussions. It is not uncommon to use free-hand details to give the information needed for coordination. After being approved, these freehand details are often drawn with drafting instruments as details that are placed on the working drawings. Construction technicians and workmen use the details to see how the various systems were designed and to choose equipment that will satisfy the requirements of the building.

This chapter will discuss details that are commonly used in the building construction industry. In earlier chapters, the design responsibilities were broken down into architectural (designing the building) and engineering (mechanical, plumbing, electrical and structural) areas. Site work plans and structural plans are included under the major heading of architectural. We will discuss the site plan section and the structural plan section under separate headings rather than grouping them under the architectural (major) heading.

Each of these areas has a need for details. Most of these details are drawn to scale; however, isometric and oblique drawings, which are not drawn to scale, are often used. This chapter will discuss information to be shown in a detail and will describe how details should be arranged on the working drawings.

23.2 WHAT DETAILS SHOULD BE DRAWN

The architects and the engineers, along with their designers/drafters, must decide what information is needed and what procedures must be followed to install the equipment. Decisions must also include coordination with other construction areas to verify that no equipment will conflict with equipment from another trade and that work done by one trade will not interfere with that done by another trade. Decisions as to what details are required are usually made during the design process.

If the drawing is too small for the required information to be shown on the floor plan, a detail of the area in question is added, enlarging a small portion of the drawing and relaying the information to the builder.

23.3   SITE DETAILS

The site plan section sometimes consists of demolition plans plus the new grading plan and the new work to be done on the site. Some details that are usually needed are as follows:

1. Storm drainage pipe installation detail

2. Catch basin details

3. Manhole details

4. Roadway paving details

5. Curb and gutter details

6. Details for planting plants and trees

7. Details of specialty items such as fountains, pools, and planters

8. Walls (brick, stone, wood, etc.)

Any or all of these details are required to complete the site work.

23.4   STRUCTURAL DETAILS

Structural details are necessary to show how the structural members are joined in order to establish safe and acceptable joints. Some of the structural details are as follows:

1. Joining of major structural steel members

2. How smaller structural members join the major structural members

3. How structural members are connected to building members

4. How structural members join footings

Structural details should include notes and dimensions so that the installer knows how to make the installation.

23.5   ARCHITECTURAL DETAILS

Some details that are commonly used in the architect's office are the following:

1. Window and window trim

2. Cornices and moldings

3. Fireplace and mantel trim

4. Doors and door trim

5. Bookcases and shelving

6. Built-in cabinets

7. Stairs and trim

The architectural detail should include such information as what types of materials are used, how different members are joined together, necessary dimensions for different members, and finishes for each member.

23.6   MECHANICAL DETAILS

Some details that are commonly used in the mechanical engineer's office are as follows:

1. Mechanical unit details

2. Fan details

3. Equipment mounting bases

4. Unit hangers details

5. Roof-mounted equipment

6. Supply and return air terminals

7. Detail of special valves

8. Duct connections

9. Piping details

10. Pump details

11. Damper details

12. Insulation details

13. Coil mounting details

14. Louver details

15. Equipment connection details

Mechanical details should include such information as the type of materials used, how different members are joined together, necessary dimensions for different members, connection points, and hangers or mounting devices.

23.7 PLUMBING DETAILS

Plumbing details, like mechanical details, are often repeated from job to job. The plumbing engineer and the plumbing technician must decide what details are needed to relay the necessary information. Some of the commonly used details in the plumbing section are as follows:

1. Interior and exterior clean-out details

2. Water heater piping detail

3. Roof drain details

4. Floor drain details

5. Piping diagrams (details)

6. Riser diagrams (details)

7. Yard hydrant details

8. Manhole details

9. Equipment connection details

Like the site, structural, architectural, and mechanical details, the plumbing details should contain all the information needed to install the detailed equipment properly. Dimensions may be required, and if so, they should be accurate and complete.

23.8 ELECTRICAL DETAILS

The electrical section must have details. The engineer and design technician must decide what details are required. Some of the commonly used details are as follows:

1. Detail for mounting and connecting special equipment

2. Electrical panel details

3. Special wiring diagrams (details)

4. Power service diagrams (weatherheads, etc.)

5. Mounting pads for electrical equipment

6. Installation details for generators

7. Mounting details for parking lot lighting

8. Details of various lighting fixtures

The electrical plans are more diagrammatic than the other plans. For this reason, the electrical section may require more details to relay information.

Like other details, the electrical details may require accurate and complete dimensions.

23.9 LOCATING DETAILS ON THE SHEET

For smaller buildings, details are usually located on floor plan sheets or other sheets where there is space available. The details are usually identified within the titles. They should be indexed in some manner so that they can be found easily. On larger jobs, details are titled and numbered for identification and are shown on a separate sheet.

23.10 DRAWING DETAILS

After a decision has been made as to the type of detail, the drafter proceeds with the drawing. Images are drawn with fine lines that are easily erased. When the drawing has been developed with light lines, a soft pencil with a fairly dull point (No. 2, for example) is used to draw over the construction lines, showing the object with crisp construction lines. Extension and dimension lines are often shown, giving the exact dimensions that are required.

SUMMARY

• Details help architects, engineers, and designers/drafters coordinate designs with the various tradespersons.

• Details sometimes start as freehand sketches that are placed on the working drawings after coordination conferences.

• Site details include demolition plans and grading and site work details.

• Structural details show how structural members are to be joined and usually include notes and dimensions.

• Architectural details list types of materials, how members are joined, and dimensions and finishes for members.

• Mechanical details list types of materials, how members are joined, dimensions, connections, and mounting devices.

• Plumbing details include information needed to install and give sizes of equipment.

• Electrical details contain diagrams to relay information and include dimensions.

• Details for small buildings are often placed on the floor plan sheets or wherever space is available; for large buildings, details are usually titled, numbered, and placed on a separate sheet.

REVIEW QUESTIONS

1. How do details help the various persons working on a job to coordinate their designs?

2. When is it decided what details should be drawn?

3. List three types of details often found in the site plan section.

4. List three types of details often found in the structural section.

5. List three types of details often found in the architectural section.

6. List three types of details often found in the mechanical section.

7. List three types of details often found in the plumbing section.

8. List three types of details often found in the electrical section.

9. Where are details usually located if the building is small?

10. Where are details usually located if the building is large? How are details identified on the drawings?

11. Discuss the steps a drafter follows in proceeding with a drawing of a detail.

24

FREEHAND SKETCHING

OBJECTIVES

After studying this chapter, you should be able to:

- describe three types of drawings that often begin as freehand sketches
- demonstrate the proper way to use guidelines for lettering notes on a plan
- demonstrate at least three factors to consider when locating information on a drawing
- list at least three factors to consider when locating notes on a plan
- demonstrate the procedure for shading a drawing to show contrast between the different parts or planes
- demonstrate how to shade an isometric or an oblique drawing to help create a three-dimensional effect

24.1 INTRODUCTION

The purpose of this chapter is to make the reader aware of the various aspects of freehand sketching techniques. No two people are alike. Everyone has natural skills that can be developed. Some people have the ability to do beautiful freehand sketching while others must struggle to become proficient. In this chapter you will learn acceptable methods for developing the needed skills for freehand sketching.

After learning the various techniques, you need to practice, practice, practice. Freehand sketching skills are learned by doing, and the information in this chapter is only an introduction. The student becomes proficient by learning new techniques and improving his or her skills each day.

24.2 TYPES OF DRAWINGS CREATED BY FREEHAND SKETCHING

Orthographic projection is probably the most common type of drawing done with freehand sketching. Orthographic projection is discussed in detail in Chapter 16. This type of drawing consists of three basic views (plan, front and side views). The front and side views are sometimes called elevation views.

In the HVAC field, the technician has many opportunities to use freehand sketching to produce plans of existing conditions. These plans are then taken back to the office where a drafter creates a scale drawing of the space. Elevation drawings can be drawn freehand and used to produce scaled drawings at a later date.

Freehand sketching is also used for making oblique drawings. Oblique drawings created with drafting instruments are discussed in detail in Chapter 17. The same procedures used for instrument drawings can be applied to produce freehand oblique drawings.

Isometric drawings are a third type of freehand sketches. Isometric drawings are drawn using the same procedures described in Chapter 18.

24.3 FREEHAND SKETCHES ARE NOT DRAWN TO SCALE

In freehand sketching, the drawings are not drawn to scale; however, these drawings must be kept in proportion to reflect the relationship of one part of each drawing to another and to show accurately the difference in sizes.

The architect's scale is often used to help the freehand drafter keep the different components of the freehand drawing in proportion. Once an approximate scale has been selected, the architect's scale can be used to establish dimensions for the drawing.

24.4 LINES USED FOR FREEHAND SKETCHING

Many types and weights of lines are drawn on freehand sketches. Refer to Chapter 15 for a discussion of the various types of lines used on working drawings. Figure 15.1, page 83, describes commonly used sizes of drafting pens and the line weights they produce. Figure 15.2, page 84, is a schedule of lines containing names, descriptions, and examples of the various types of lines found on drawings. Freehand sketches use the same types and weights of lines as working drawings.

24.5 LETTERING FOR FREEHAND SKETCHING

The beginning drafter soon learns that the lettering of notes on a drawing is very important in freehand sketching. The same lettering and notation procedures that were discussed in Chapter 19 are used in freehand sketching and drafting.

The more proficient the drafter is at lettering, the better the drawings look. See Chapter 19 for information on forming letters properly for freehand sketching and for manual drafting.

Guidelines should always be used for hand-lettering a plan (see Chapter 19). These lines help the drafter to make the letters the same size and properly aligned.

Technicians often develop special flairs in their systems of lettering. Even with basic uppercase letters, drafters can achieve a unique appearance by developing variations of the standard block or the standard sloped-block letters. A few examples of these modifications are shown in Figure 24.1.

24.6 LOCATING NOTES ON FREEHAND SKETCHES

Notes on drawings are very important. Freehand sketches and drawings are noted in the same manner as other drawings (see Chapter 19).

In most cases the drafter or technician completes the drawing before lettering the required notes. The organization and location of notes should be planned in advance of putting them on the plan. How to add notes to a freehand sketch is discussed in Chapter 19.

24.7 LOCATING DRAWINGS ON A PAGE

Planning prior to starting the actual drawing process is very important. The sheet must be neat and well-organized so that it is easily read. The drawings should not be crowded or placed in spaces that are not suitable. To be sure that an object or building will fit on the sheet, it should be drawn to an approximate scale and then located on the sheet.

To locate the object being drawn in the middle of the sheet, follow these steps: Locate the center of the object (or building) based on its dimensions. Find the center of the sheet, and then place the center of the object on the center of the sheet.

ENGINEERS AND ENGINEERING TECHNICIANS USE SLOPED LETTERS IN THE MAJORITY OF CASES.

SOME DRAFTERS LIKE VERTICAL LETTERS.

SOME DRAFTERS USE A TRIANGLE TO FORM ALL THE VERTICAL PORTIONS OF THE LETTERS

SOME DRAFTERS OVERSHOOT GUIDE LINES WITH SUCH LETTERS AS F, H, L, P, T AND Y.

Figure 24.1 Lettering

24.8 DIMENSION AND EXTENSION LINES

As discussed in Chapter 11, construction drawings cannot be scaled for accurate dimensions because of the stretching or shrinking that takes place as copies are made of the working drawings. It is the responsibility of the drafter to write the dimensions on working drawings so that the dimensions can be read directly from the plan. The freehand drafter should show all important dimensions on the freehand sketch so that the correct dimensions are available to the drafter. Chapters 15 and 25 describe ways to show dimensions on the plan.

24.9 SHADING AND CROSSHATCHING FREEHAND DRAWINGS

Shading and crosshatching are discussed in detail in Chapter 21. Shading and crosshatching pertaining to freehand sketching will be briefly discussed here.

The term *shading* has several meanings. One meaning is "the act of 'coloring', crosshatching or adding special markings to various surfaces of an object to make these surfaces stand out from the remainder of the drawing." Some of this special treatment of surfaces is shown in Figure 24.2. Figure 24.3 is an example of crosshatching, and Figure 24.4 shows special treatment to represent how the actual material looks in an air-handling unit. Poché (pronounced "poshay") is commonly used by architects to describe special treatment (see Chapter 21). In Figure 24.5, an example of poché is shown.

A special transparent sheet with a series of dots or dashes is available for shading surfaces. This transparent material has a sticky backing that adheres to the working drawing. An architect or engineer uses this manufactured material because it is easy to use and is a good, consistent way to shade or highlight surfaces on working drawings.

The second variation of shading is by highlighting surfaces in isometric or oblique drawings to make them look more realistic and to achieve a three-dimensional appearance. Even though the drawing is shown on a single plane, it has a three-dimensional effect. Examples of shading by highlighting are shown in Figure 24.6.

Figure 24.2 Special treatment of surfaces (freehand)

To shade an isometric or oblique drawing, the drafter should visualize how the object would look with a spotlight trained on the object from an angle. The irregular shapes would cast shadows on the object where the rays of light did not strike the surfaces. In shading, the coloring or shading is placed on the object on the same surfaces where shadows would be cast if the object were seen in three dimensions. If the shading is done properly, the effect is the same as if the object were seen naturally with the spotlight being directed from an angle onto the object. Several examples of shading isometric and orthographic drawings are shown in Figures 17.4, 17.5, and 17.6, page 106, and Figures 21.7 and 21.8, page 137.

Figure 24.3 Freehand drawing showing different surfaces crosshatched

Figure 24.4 Crosshatching various components of an air-handling unit

Figure 24.5 Architectural detail with poché on wood studs

Figure 24.6 Highlighting shading to improve drawing

SUMMARY

- Freehand sketching skills are learned by practicing learned techniques.

- Orthographic projection, oblique drawings, and isometric drawings can be drawn with freehand sketching.

- Freehand sketches are not drawn to scale, but must be kept in proportion to show accurate relationships of the various parts of the drawing.

- Guidelines are used in hand-lettering a plan to keep letters in a straight line and about the same size.

- Planning prior to starting the actual drawing process is very important.

- Dimensions should be written on the drawings and sketches since construction drawings cannot be scaled for accurate dimensions.

- Poché, shading, and crosshatching make various surfaces stand out on the object being drawn.

REVIEW QUESTIONS

1. What are three types of drawings that often begin as freehand sketches by the technician in the field?

2. Why must freehand sketches be in proportion even though they are not drawn to scale?

3. What is the best way to learn freehand sketching skills?

4. Why is it important to write dimensions on the plans instead of relying on reading the scale?

5. Which is usually done first, the drawing or the notes?

6. What are some of the steps to take when locating an object or building on a sheet?

7. Why does the drafter often use a commercially prepared transparent paper for shading?

8. To shade an isometric or oblique drawing, how does the drafter visualize the object?

ADDITIONAL STUDENT EXERCISES: Freehand Sketching

1. Within the confines of the enclosed space (Figure 24.7), freehand sketch an equipment room containing a boiler, air-handling unit, and a round water heater. Use rectangles to represent the boiler and air-handling unit. The air-handling unit is to be approximately one-third larger than the boiler.

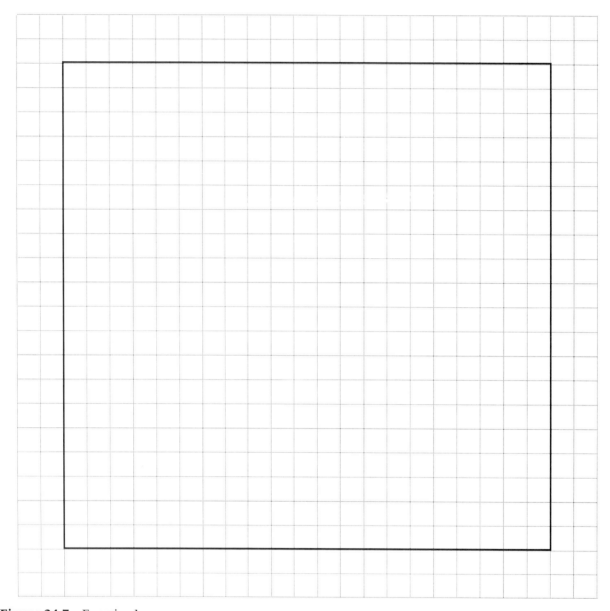

Figure 24.7 Exercise 1

2. In the space provided in Figure 24.8, freehand sketch an isometric drawing of an outside condensing unit, a cooling coil, and a round water heater. (Turn sheet to 90° to have graph paper lines accommodate isometric drawings.) Shade units to improve appearance of drawings.

Figure 24.8 Exercise 2

25

DRAFTING WITH INSTRUMENTS

OBJECTIVES

After studying this chapter, you should be able to:

- discuss reasons why design, accuracy, and speed affect the success of a project

- identify commonly used manual drafting instruments

- describe and identify time-saving procedures for drafting

- discuss and demonstrate ways to draw air-conditioning components

- read and explain a simplified plan for an equipment room

- demonstrate the ability to draw ductwork and fittings with manual drafting instruments according to directions

25.1 INTRODUCTION

The success or failure of a building construction project often can be determined by the quality of the construction plans. If the plans are correctly drawn with no dimensional mistakes and if all the elevations, sections, details, and schedules are correct, the construction personnel will have little trouble with construction. If the plans have drafting errors, dimensional mistakes, or if the drawings cannot be read, the builder will have to spend precious construction time in trying to find correct answers

or in tearing out work that is not correct. Even at best mistakes will be made; however, these mistakes can be held to a minimum if drawings are checked and rechecked for accuracy and clarity.

For the successful completion of any project, architects, engineers, designers, and drafters are constantly concerned with three factors: design, accuracy, and speed. This chapter will concentrate on the accuracy, good design, neatness, drafting speed, and common sense needed on the plans to relay the necessary information to the builder. Manual drafting instruments will be described, and drafting procedures will be illustrated.

25.2 COMMONLY USED MANUAL DRAFTING INSTRUMENTS

Years ago all working drawings were produced using drawing boards and T-squares. Refer to Figure 8.1 of this textbook for an illustration of a drawing board and T-square along with an illustration of a drawing board with a drafting machine and parallel bar. These devices are widely used in drafting rooms today to establish horizontal lines. With triangles placed on the T-square and parallel bar, vertical lines and angular lines can be drawn. The drafting machine shown in Figure 8.1, page 40, is used to draw horizontal lines and vertical lines (90 degrees off horizontal), along with any angular line in the 360-degree field. The angular lines can be adjusted with the drafting machine head. Once the angular line has been selected, the head is locked into place so that the angular line can be drawn without the use of triangles.

Additional manual drawing instruments shown in Figure 8.2, are as follows:

1. Dividers (often called "points") — used to transfer dimensions or distances on drawings.

2. Compass — has one sharp metal point and one graphite lead point. The compass is used to draw circles and arcs.

3. Triangles — usually come in pairs (a 30/60-degree triangle and a 45-degree triangle). These triangles, used with the T-square or the parallel bar, are used to draw vertical lines and angular lines.

4. Protractor — used to measure angles (see Figure 4.9, page 20).

5. Plastic template — sheet of plastic with cut-outs of circles, squares, triangles, and other geometric figures allowing the drafter to insert a pencil or pen in the cutout and draw the figure on the plans. Using the template saves drafting time.

6. Architectural scale and engineering scale — used to reduce the size of a drawing (see Chapters 5 and 6).

7. Drafting pencils (wood and mechanical) and drafting pen — used to render drawing.

Other manual drafting equipment includes erasing shields and lettering guides. Erasing shields are thin sheets of metal with cut-outs (Figure 25.1). For erasing, the shield is placed over the portion of the drawing that is not to be erased. Through the cut-out, the unprotected part of the drawing

can be erased without disturbing the remainder of the drawing. Lettering guides are used to draw guidelines so that letters will have the same height. (See Chapter 19 for a discussion of lettering guides.)

25.3 DRAWING LINES ON WORKING DRAWINGS

In drafting, the use of different kinds of lines helps the blueprint reader read and interpret the drawing. These lines should be drawn straight and should end abruptly. When two lines intersect at a corner, they should end at the corner and not overshoot or leave a gap at the intersection (Figure 25.2). The quality of the lines throughout the drawing should be uniform. All object lines should be the same, and all hidden lines, center lines, extension and dimension lines, leaders, and so forth should be the same at all locations on the drawing.

Chapter 15 is devoted to drafting lines. Figure 15.1, page 83, shows common line weights (width and darkness), and Figure 15.2, page 84, is a line schedule, or alphabet of lines. Other figures in Chapter 15 illustrate the use of different types of lines.

All circles and arcs should be drawn with uniform lines, and the circle should be complete with no starting and ending points (Figure 25.3). Arcs should have definite starting and ending points.

25.4 TIME-SAVING DRAFTING PROCEDURES

With experience, the drafter learns shortcuts that save both time and effort. Figures 25.4 through 25.7 demonstrate some of these time-saving procedures.

Figure 25.4 shows how to bisect an angle using a compass. Figure 25.5 illustrates how to locate the center of a circle if the center is unknown. Figure 25.6 shows how to divide a line segment into equal

Figure 25.1 Erasing shield

Figure 25.2. Intersection of lines

CORRECT   INCORRECT   INCORRECT

Figure 25.3 Drawing a circle

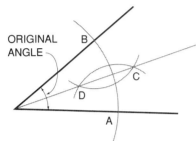

ORIGINAL
ANGLE

NOTE: (1) WITH A COMPASS, SWING AN ARC WITH POINT OF
COMPASS ON POINT OF ANGLE. (2) SWING AN ARC WITH
COMPASS POINT AT POINT "A." SWING THE SAME ARC WITH
COMPASS POINT AT "B." DRAW A CONSTRUCTION LINE BETWEEN
POINTS "C," "D" AND THE POINT OF THE ANGLE. THIS
CONSTRUCTION LINE DIVIDES THE ORIGINAL INTO TWO
EQUAL PARTS.

Figure 25.4 Bisecting an angle

NOTE: DRAW A SQUARE
AROUND THE CIRCLE. DRAW
CONSTRUCTION LINES
DIAGONALLY FROM
CORNERS. WHERE THESE
TWO LINES CROSS IS THE
CENTER OF THE CIRCLE.

Figure 25.5 Finding the center of a circle

parts. The example shown divides the line segment into twelve equal parts, but any number of parts can be found by changing the angle of the scale or by changing the scale.

A circle can be divided into equal segments as shown in Figure 25.7. First, the estimated distance is set on the divider or compass. The divider or compass is "walked" around a quarter of the circle. If the distance is incorrect, the divider or compass is adjusted up or down (by estimation), and the instrument is again "walked" around the quarter of the circle. This procedure is repeated until the

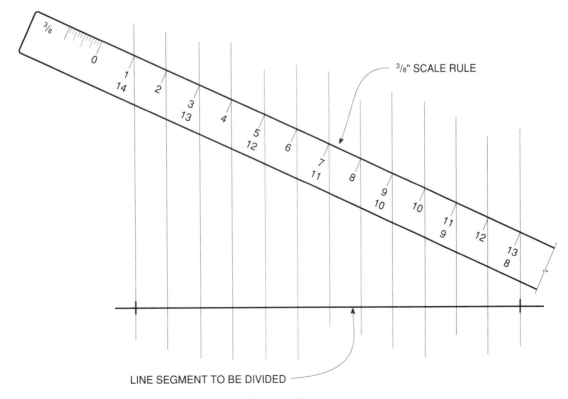

LINE SEGMENT TO BE DIVIDED

Figure 25.6 Dividing a line segment into twelve equal segments

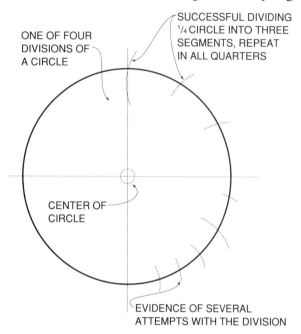

Figure 25.7 Dividing a circle into twelve equal segments

distance is determined to be correct. To check the accuracy of the determined distance, the divider or compass is then "walked" around the entire circle. This procedure can also be used to divide line segments into equal segments.

25.5 DRAWING HEATING AND AIR-CONDITIONING COMPONENTS

The ductwork elbow is one of the most common fittings used in a duct system. Figure 25.8 shows a pictorial view of a 90-degree elbow with the various parts of the elbow identified. A 45-degree and a 30-degree elbow have the same name as a 90-degree elbow. Each side of the elbow is called a "cheek". The short radius on the inside of the ell is called the "throat," and the outside radius called the "heel," or "wrapper".

The procedure for drawing the elbow is shown in Figure 25.9. The throat radius (R-1) is usually one-half the width of the ductwork with a minimum of six inches. The R-1, or "throat," radius is drawn to connect the two sections of ductwork. The larger radius (R-2 or "heel") is drawn to connect the outside of the elbow. In some cases a square throat is required (Figure 25.10).

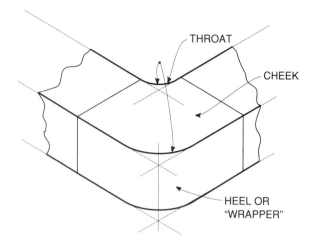

Figure 25.8 Parts of a 90° square or rectangular elbow

Figure 25.9 Drawing a 90° elbow

Figure 25.10 Round elbow with square throat

A splitter fitting for a rectangular duct system is shown in Figure 25.11. Drawing a splitter fitting is very similar to drawing an elbow. The throat radius (R-1) is established, and the main duct is divided into segments representing the quantity of air to be supplied in each duct. The large radius (R-2) is drawn from this division point to complete the "heel" portion of the take-off.

The procedure for drawing the single splitter fitting is shown in Figure 25.11. A three-way splitter damper is illustrated in Figure 25.12. In both cases, the main duct is divided to represent the percentage of air to be supplied in each branch. The inside radius (R-1) is drawn, and the larger radius (R-2) is drawn to connect the divided segment of the main duct into the various segments downstream to the main duct.

The drawing procedure for a ductwork offset is shown in Figure 25.13. Here again, the small radius (R-1) is usually set at six inches, and the large radius (R-2) is set as required to accommodate the duct size without restricting the air flow. A straight line is used to connect the arcs and to complete the fitting.

The ductwork transition can reduce the size on one side, both sides (central transition), bottom, or top. The side shown on the plan (plan view or elevation view) shows how the transition is made on that side, but the other transitions must be noted (FOT= flat on top, FOB = flat on bottom).See Figure 25.14 for an example of ductwork transition that is flat on top.

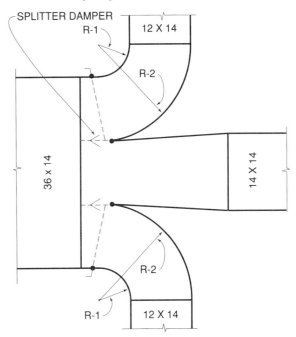

Figure 25.12 Three-way splitter damper

25.6   EQUIPMENT ROOM PLAN

Figure 25.15, page 164, is an example of a simple mechanical equipment room with a boiler, air-handling unit, hot water circulating pump, expansion tank, and hot water piping. Although the plan does not show all details that are required to complete the system, enough information is given to locate the major equipment. The general location of the hot water piping within the equipment room is given. The ductwork has been shown, along with major fittings and duct sizes.

Major dimensions are given so that the equipment can be located in the space. Notice that service space has been provided around each unit. Housekeeping pads have been shown under each piece of equipment, and floor drains (installed by the plumbing

Figure 25.11 Single splitter fitting

R-1 EQUALS ONE-HALF THE WIDTH OF THE DUCTS.
R-2 EQUALS R-1 PLUS THE WIDTH OF THE DUCT.

Figure 25.13 Drawing an off-set fitting

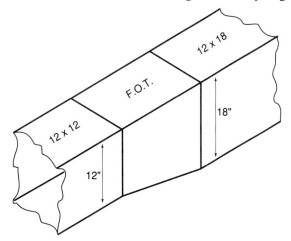

Figure 25.14 Ductwork transition that is flat on top

contractor) have been shown so that piped drain piping can be extended from the equipment by the mechanical contractor.

The drafter will be required to draw sections and details of these pieces of equipment to show specialties such as the safety valve and control devices on the boiler, the cold water makeup for the boiler, the hanging of the expansion tank, the boiler fitting, and piping connections to the boiler.

The piping connection to the pump should be detailed so that the valving, method for mounting, and other details can be shown.

The air-handling unit should have details to show how the hot water piping is connected to the heating coil. A separate detail of the heating coil may be

necessary to show the coil connections that would include the controls, bypass, valves, and so forth.

The air-handling unit would normally have a cooling coil and chilled water or refrigerant piping. For clarity in the description of the drawing, this cooling system has been omitted.

All lines on this plan were drawn with the drawing instruments described in Section 25.2 and illustrated in Chapter 8. All of the straight lines were drawn with a parallel bar and triangles. Circles representing piping elbows turned up and down were drawn with a circle template. The arcs necessary to draw the ductwork elbows were drawn with a circle template and with a compass.

SUMMARY

• Working drawings can be drawn manually with a T-square and drawing board, or with a drafting machine and parallel bar.

• Other manual drawing instruments include dividers, compasses, triangles, protractors, templates, erasing shields, and lettering guides.

• Lines on drawings should conform to a standard Table of Line Weights and Alphabet of Lines.

• Experienced drafters learn time-saving procedures using compasses, dividers, and other manual drafting instruments.

• The drafter can gain valuable experience by manually drawing the various components of a heating and air-conditioning system.

REVIEW QUESTIONS

1. Define and describe the following drafting instruments:

   A. T-square

   B. Drafting machine

   C. Parallel bar

   D. Dividers

   E. Compass

   F. Triangles

   G. Template

   H. Scale rules

   I. Drafting pencil

   J. Erasing shield

   K. Lettering guide

2. Formulate directions for the following procedures, and demonstrate each with an example:

   A. Drawing straight lines that intersect at a corner

   B. Drawing a circle

   C. Bisecting an angle using a compass

   D. Dividing a line segment into several parts

   E. Dividing a circle into equal segments with a compass

3. Draw a ductwork elbow, and identify the parts.

4. Describe the procedure for drawing the following:

   A. Single splitter fitting

   B. Three-way splitter fitting

5. List the major components of a simple mechanical equipment room.

ADDITIONAL STUDENT EXERCISES: Drafting with Instruments

1. Refer to Figure 25.15 to answer the following questions.

   A. Is there a chimney on this drawing?

   B. Is there an expansion tank?

   C. Is there a housekeeping pad under the boiler?

   D. Are flexible duct connections shown on ductwork?

   E. Is there a splitter damper in the supply duct?

   F. Does the boiler supply hot water to the second floor?

   G. How many gate valves are shown on this drawing?

   H. Are any unions shown?

   I. Is the ductwork located underfloor?

Figure 25.15 Equipment room with boiler, air-handling unit, hot water pump, expansion tank, ductwork, and hot water piping

2. (A) Draw a 90-degree elbow on the end of the indicated duct in Figure 25.16A. The elbow is to have a round "throat" and a round "heel."

(B) Figure 25.16B is an offset section of duct work. Swing the minimum and the maximum arcs, and connect these arcs with a straight line.

(C) Using Figure 25.16C, complete the transition that is to be flat on top.

(A) 90° DEGREE, ROUND THROAT ELBOW. THROAT RADIUS IS EQUAL TO 1/2 THE WIDTH OF DUCT.

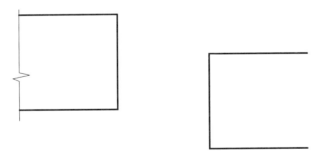

(B) OFFSET WITH THE SMALL RADIUS EQUAL TO 1/2 THE WIDTH OF DUCT.

(C) DOUBLE TRANSITION WITH THE DUCT FLAT ON TOP.

Figure 25.16A-C   Exercise 2

3. Refer to Figure 25.17. Draw the indicated 90-degree elbows, the splitter fitting, the offset, and the transition.

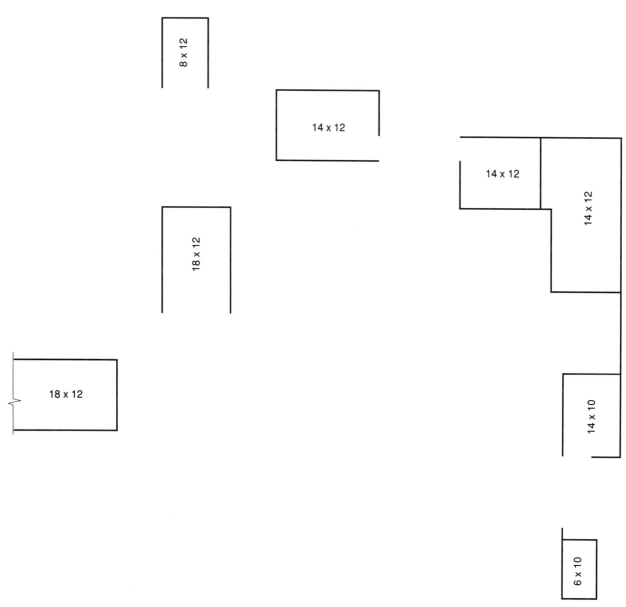

Figure 25.17 Exercise 3

4. Refer to Figure 25.18. Use the symbols shown in Appendix D to draw the piping connections to the heating coil. Connect the heating coil with a three-way valve in the H.W. "OUT" line. Show a union in the hot water connection "IN" and "OUT." The bypass line off the three-way valve shall have a union and a ball valve for balancing flow. Show a gate valve in the "OUT" line, upstream of the three-way valve. Install a globe valve in the H.W. line upstream of the bypass line of the three-way valve.

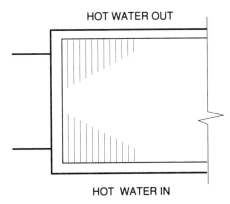

HOT WATER OUT

HOT WATER IN

Figure 25.18   Exercise 4

PART III

READING AND INTERPRETING ARCHITECTURAL PLANS

26

READING ARCHITECTURAL PLANS

OBJECTIVES

After studying this chapter, you should be able to:

- identify parts of the architectural floor plan
- describe the information found on the roof framing plan.
- locate and discuss the importance of the transverse and longitudinal sections
- locate and discuss the importance of the wall sections and details

NOTE: Refer to the plans for "A Small Bank and Trust Company" (at the end of this textbook) while studying this chapter.

26.1 INTRODUCTION

The HVAC technician needs to be able to read and interpret complete architectural plans. In reviewing these plans, it is suggested that each sheet be read similar to the way one reads a book. The technician should visualize the sheet in squares of approximately six inches and then read each section from left to right for a general idea of the information in that part. When all parts of the sheet have been reviewed in this manner, the technician will have a general idea of what information is given on the sheet, what the building looks like, and what systems have been designed for the building. In this chapter, students are given an opportunity to look at actual architectural plans to find various features, details, building materials, and construction procedures for a building.

26.2 SITE PLAN AND LOCATION PLAN

The site plan and location plan are shown on Sheet 1 of the plans at the end of the book. The larger drawing is the location plan. This plan gives the location of the project by listing the U.S. highway and state route intersection. (The highway and route numbers have been omitted.) This plan also shows the proposed shopping center to be built on the property as well as the present and future drives.

The site plan shows the building located on the site along with the drives, parking, planters, and so forth. Contour lines are shown along with the bench mark. The bench mark is a known elevation point, usually a concrete pad or a metal stake driven in the ground. This bench mark determines all elevations needed in the grading of the site and all elevations to be used for the building.

26.3 FOUNDATION PLAN

The foundation plan shows the foundation for the building. Dimensions are given for the outside walls of the building, and sections are indicated so that the builder will know how to install the building foundation. This plan is critical because if the foundation is not properly installed, the building walls may fail. If the dimensions are not correct, the building may not fit the requirements of the plan.

26.4 ARCHITECTURAL FLOOR PLAN

The architectural floor plan is shown on Sheet A-1 at the end of this textbook. This sheet contains information about the various features of the building such as walls, windows, doors, floors, overhangs, types of walls, plumbing fixture locations, and so on. The HVAC design technician must be familiar with all of these items in order to design a suitable system. The installing technician must know how the building is put together and where the various items are to be located for maximum efficiency.

On Sheet A-1, sectional drawings show where sections have been cut through the walls. The sections are indicated by circles with arrows pointing in the direction that the section is cut. For example, there are three sections through the right-hand wall and several sections through the front wall. These are shown on Sheets A-4 and A-5 and will be discussed later in this chapter.

26.5 WALL SECTIONS

On Sheet A-2, wall sections are cut to allow elevations to be shown on Sheets A-4 and A-5. On the elevations sheets, the HVAC technician can measure the square footage of wall spaces for windows, doors, overhangs, and other features of the building that affect the heating and air-conditioning load. On the front and rear elevations of this set of plans, a note under the title indicates that the front and rear elevations are the same except reversed. This is true of many small buildings such as small bank buildings. The front and rear are the same, but the two side elevations are different, as can be seen by examining the plans.

26.6 ROOF FRAMING PLAN

Sheet A-3 shows the roof framing plan. This is the structural plan. The sizes of steel members are

given, along with the location of steel joists. This information is necessary for the person designing the system, who needs to know whether to hang the ductwork under the steel joists or in between them. The joists are the steel structural members that support the roof. In some cases, air-conditioning units are located on the roof, and steel must be sized large enough to carry the extra weight of the air-conditioning units. The sections shown through the building are located on Sheet A-5.

26.7 TRANSVERSE AND LONGITUDINAL SECTIONS

The transverse sections and longitudinal sections give a cross section of the building as it will be constructed. These sections show what the steel looks like and what spaces are available for ductwork and piping. The transverse and longitudinal sections also give ceiling heights, construction and location of beams, and location of the roof floor. They show exactly what the building looks like as it is cut along the cutting plane. These transverse and longitudinal sections are a very important part of the architectural plan.

26.8 WALL SECTIONS AND DETAILS

Sheet A-6 is titled Wall Sections and Details. These sections, which have been referred to on other sheets, show how the various components are put together to form the building. Sheet A-6 gives elevations and dimensions on a larger scale so that the builder knows exactly how the building is to go together. The HVAC technician, whether designing the system or installing it, uses these parts of the plan to decide where to put ductwork and how to install it properly.

26.9 ENLARGED FLOOR PLANS AND INTERIOR ELEVATIONS

An enlarged floor plan and elevation of the tellers' area are shown on Sheet A-7. This sheet shows the design of the tellers' stations and includes details and dimensions for the cabinetmakers. This drawing helps the HVAC contractor design the heating/cooling system that will best serve the tellers.

26.10 SCHEDULES, SECTIONS, AND ELEVATIONS

The Room Finish Schedule, the Door Schedule, and the Color Product Schedule are shown on Sheet A-8. The partial plan for the public lobby glass wall and doors is shown along with the elevation of the wall. Other details for other parts of the lobby wall are also shown.

SUMMARY

- The location plan gives details as to the intersection where the building is to be located.

- The site plan shows how the building is to be placed on the site, along with drives, parking areas, planters, and other features.

- The bench mark is an elevation point to which all elevations are referenced.

- The foundation plan shows dimensions of the building with sections that show how the foundation is to be built.

- The architectural floor plan shows various features of the building and includes sectional drawings.

- Wall sections give dimensions of windows, doors, and other features of the building.

- The roof framing plan gives information on the structural requirements of the building.

- Transverse and longitudinal sections give cross sections of the building at various locations.

- Schedules give specific information on doors, colors, and other details of the building.

REVIEW QUESTIONS

1. Describe the method suggested in Section 26.1 for getting an overview of the plans.

2. What kind of information is given on the location plan?

3. What kind of information is given on the site plan?

4. What types of information are given on the wall sections?

5. Why is the information on the roof framing plan especially important to the HVAC technician?

6. Describe the kinds of information given on the transverse and longitudinal sections.

7. What are some other parts of the plans that furnish important information to the builders and contractors?

ADDITIONAL STUDENT EXERCISES: Reading Architectural Plans

A. THE SITE PLAN—SHEET 1

Using the set of building plans at the end of the book, answer the following questions or perform the described tasks for the site plan (Sheet 1).

1. Which is closer to the bank building, the U.S. highway or the state route?

2. On which side of the building is the drive-up window located? (north, south, east, or west)

3. Have parking spaces have been provided for customers of the bank?

   4. What kind of walkways lead to the bank entrances?

   5. What is the width of the power line easement?

   6. What is the finished floor elevation of the building?

   7. Is there any existing pavement on the site other than the road and highway?

   8. Does the shopping center exist, or is it proposed?

   9. Is there an exit/entrance on the north side of the building?

   10. What is the size of the existing water main?

   11. Is the size of the existing sanitary main given?

   12. Are standard symbols used for the domestic water main and the sanitary water main?

   13. Draw an example of the standard symbols for the domestic water main.

   14. Draw an example of the standard symbol for the sanitary sewer main.

   15. How many sanitary manholes are shown on the site plan?

   16. Is the sanitary sewage pumped up to the sanitary sewer main?

   17. How deep should the concrete curb be below the finished pavement?

   18. Where is the sewage lift pump located?

   19. How many brick panels are shown for each walkway to the building?

   20. What is the elevation of the bench mark?

   21. How big is the plot of land that the bank is to be built on?

   22. Who furnished information about the site to the architect?

   23. What is the elevation of the drive-up window pavement?

   24. How much higher is the bank floor than the drive-up window pavement?

   25. Can any part of the building be located on the power line easement?

B.   THE FOUNDATION PLAN—SHEET 2

Using the set of building plans at the end of the book, answer the following questions or perform the described tasks for the foundation plan (Sheet 2).

   1. What section number shows the column located near the center of the building? (Find the section drawing, and list the numbers in the top and bottom half of the circle that indicates that section.)

   2. What is the size of the concrete footing for the column near the center of the building?

   3. What do the broken lines shown on each side of the foundation wall represent?

   4. Do the columns supporting the canopy over the drive-up window have the same type of footing as the column inside the building?

   5. What type of construction is used in the vault floor?

   6. How thick is the vault floor?

   7. What is the size of reinforcement bars used in the construction of the vault?

   8. What is the door that is shown inside the vault used for?

9. Is carpeting shown in the vault?

10. What is the size of the concrete footings under the vault?

11. How many reinforcement bars are required in the vault footings, and what size are they?

12. What is the wall construction around the drive-up window?

13. What is the wall construction around the depository?

14. How thick is the wall around the depository?

15. Why is the wall under the depository reduced to 8" thick?

16. What material is used for foundation walls?

C. THE FLOOR PLAN—SHEET A-1

1. How many teller stations are shown?

2. What are the inside dimensions of the conference room?

3. How much is the floor slab depressed to accommodate the brick pavers at the tellers' counter?

4. How many check desks are shown?

5. In what room is the roof access ladder located?

6. What are the dimensions of the lounge (Room 102)?

7. Does the kitchen unit have a sink?

8. Does the building have two entrances?

9. What scale was used to draw the original working drawing? (Remember that the plans in your textbook have been reduced, and the scale is no longer valid.)

10. What scale was used to draw the elevation in the employee lounge?

11. What scale was used to draw the detail showing the depression for brick pavers in the lobby?

12. What kind of columns are used in the building?

13. What is the overall width of the entrance?

14. How is the roof overhang shown?

D. THE ELEVATION PLAN—SHEET A-2

1. What material is used for the fascia?

2. From what material is the gravel stop fabricated?

3. What kind of glass is used?

4. How thick are the glass panels?

5. What is the special brick that is installed between the brick panels on the outside wall?

6. Looking at brick details, describe a sailor course.

7. Is the drive-up window unit a manufactured product?

8. Who installs the drive-up window unit?

E.　THE STRUCTURAL PLAN—SHEET A-3

1. What are the sizes of the two structural beams located near the center of the building?

2. What is the size of the pipe columns?

3. What is the size of the steel joists used in the center of the building?

4. How many of these joists are required?

5. What is the size of the steel joists used in the outer portion of the building?

F.　THE REFLECTED CEILING PLAN—SHEET A-4

1. What type ceiling is used in the main part of the building?

2. What material is used around the perimeter of the ceiling?

3. The round circle with double crosshatching at 90 degrees is a symbol representing speakers. How many speakers are shown on this job?

4. What is the ceiling height in the main lobby?

5. What is the ceiling height over the teller area?

6. What is the ceiling height in the storage area?

7. What is the ceiling height of the outside soffit?

8. On this sheet, what materials are shown in the outside wall section?

G.　TRANSVERSE AND LONGITUDINAL SECTIONS—SHEET A-5

1. Does the cashier's cabinet show on these sections?

2. Are hangers for the acoustical ceiling shown?

3. What scale was used to draw these sections? (Remember that the scale is no longer accurate because plans have been reduced to fit in this textbook.)

4. Is the drive-up window shown?

5. Is compacted fill shown on the plan?

H.　ARCHITECTURAL SECTIONS—SHEET A-6

1. What is the finish material on the fascia?

2. In Section 5/A-1, A-2, and A-3, what is the size of the steel beam structural member that supports the bar joists?

3. What is the size of the bar joists shown in Sections 4 and 5?

4. Are bar joists used in the soffit construction?

5. What type of insulation is shown on the roof?

6. What kind of roof deck is shown?

7. What is meant by sailor coursing of brick?

I.　ELEVATIONS—SHEET A-7

1. Has knee space been provided for the tellers?

2. Are foot rests provided?

3. How many tellers' stations are shown?

J. SCHEDULES—SHEET A-8

1. What information is shown on the Room Finish Schedule?

2. What information is given on the Product Schedule?

3. What information is given on the Door Schedule?

4. Why is an elevation of a wall, along with a plan for the wall, in the public lobby shown on this sheet?

27

DUCTWORK PLANS

OBJECTIVES

After studying this chapter, you should be able to:

- read and use a sheet metal chart to determine what gauge metal to use for various sizes of ductwork

- explain how linear measurements on ductwork and fittings are obtained

- use a ductwork take-off form correctly

- read and use a chart for measuring ductwork insulation

- discuss the various factors that determine the cost of ductwork and insulation, including the waste factor and labor costs

27.1 INTRODUCTION

For the majority of air-conditioning and heating jobs, ductwork is included as part of the installation. Several kinds of ductwork are commonly used in today's market, including conventional sheet metal ducts, fiberglass duct-board ducts, and flexible ducts. Special duct systems can be fabricated from fiberglass and other modern materials.

This chapter will discuss a method for estimating the amount of sheet metal required to fabricate ductwork in the shop for a job. The discussion will include how to take off (measure) the ductwork from a set of plans and how to estimate the number of pounds of metal required to fabricate the ductwork. The method recommended in this chapter has been proven to be satisfactory and simple to use.

The estimating of prefabricated sheet metal ducts, fiberglass duct-board ducts, and flexible duct materials will not be included in this chapter. Manufacturers of these products suggest methods for estimating such duct systems.

An approved method for estimating the amount of insulation required for a ductwork system will also be discussed in this chapter. There are several methods for estimating ductwork insulation. The method described in this chapter is a proven system that is simple to use.

27.2 WHAT KIND OF DUCTWORK IS SHOWN ON THE PLANS

The kind of ductwork to be used on the job is usually noted and described on the plans or described in the specifications. In some cases, the kind of ductwork is described in both places. One set of construction plans may show several different kinds of ductwork.

27.3 SHOP-FABRICATED SHEET METAL DUCTWORK

Large air-conditioning and heating contractors have their own sheet metal shops that produce all the ductwork used by the company.

To estimate the cost of fabricating and installing ductwork, a dependable method must be developed. Figure 27.1 is the Sheet Metal Duct Estimating

Factors Chart. This chart, properly used, helps a drafter or take-off person to estimate the cost of a duct system.

27.4   USING THE DUCT FACTOR CHART

The Duct Factor Chart (Figure 27.1) is used to aid the estimator in determining the correct metal gauge and weight per linear foot for duct sizes. Figure 27.1 has eight columns labeled "1" and eight columns labeled "2." The numbers listed in Column 1 are obtained by adding together the two dimensions of a square or rectangular duct. The numbers in Column 2 represent the weight of metal that is required to make one foot of ductwork of the size given and the gauge metal selected.

The gauge metal to use is determined by the larger dimension of the rectangular duct size. (For example, the "larger dimension" of a duct 24" x 36" is thirty-six inches.) The gauges are given in the lined-in rectangular blocks on the ductwork chart. Four different gauges are shown on the chart: 26 gauge (for larger duct size dimension of 0"–12"), 24 gauge (for larger duct dimension sizes of 13"–30"), 22 gauge (for larger duct dimension sizes of 31"–60"), and 20 gauge (for larger duct dimension sizes of 61"–90"). Metal gauges refer to the thickness of the metal. Figure 27.2 is a table of metal gauges, thicknesses, and weights per square foot.

SHEET METAL DUCT ESTIMATING FACTORS CHART

1	2	1	2	1	2	1	2	1	2	1	2	1	2	1	2
0–12" 26 GA.															
12	2.00	30	6.5	56	12.13	55	13.75	81	20.25	107	26.75	77		103	
13	2.17	31	6.72	57	12.35	56	14.0	82	20.5	108	27.0	78	22.8	104	30.4
14	2.34	32	6.94	58	12.56	57	14.25	83	20.75	109	27.25	79		105	
15	2.50	33	7.15	59	12.80	58	14.5	84	21.00	110	27.5	80	23.4	106	30.9
16	2.67	34	7.37	60	13.0	59	14.75	85	21.25	111	27.75	81		107	
17	2.84	35	7.59	**31–60" 22 GA.**		60	15.0	86	21.5	112	28.0	82	23.9	108	31.5
18	3.0	36	7.8	36	.9.0	61	15.25	87	21.75	113	28.25	83		109	
19	3.17	37	8.02	37	9.25	62	15.5	88	22.0	114	28.5	84	24.5	110	32.1
20	3.34	38	8.24	38	9.5	63	15.75	89	22.25	115	28.75	85		111	
21	3.5	39	8.45	39	9.75	64	16.0	90	22.5	116	29.0	86	25.1	112	32.7
22	3.67	40	8.67	40	10.0	65	16.25	91	22.75	117	29.25	87		113	
23	3.84	41	8.89	41	10.25	66	16.5	92	23.0	118	29.5	88	25.7	114	33.3
24	4.0	42	9.1	42	10.5	67	16.75	93	23.25	119	29.75	89		115	
13–30" 24 GA.		43	9.32	43	10.75	68	17.0	94	23.5	120	30.0	90	26.3	116	33.9
		44	9.54	44	11.0	69	17.25	95	23.75	**61–90" 20 GA.**		91		117	
19	4.11	45	9.76	45	11.25	70	17.5	96	24.0			92	26.7	118	34.4
20	4.34	46	9.97	46	11.5	71	17.75	97	24.25	67	19.66	93		119	
21	4.55	47	10.2	47	11.75	72	18.0	98	24.5	68	19.9	94	27.4	120	35.0
22	4.77	48	10.42	48	12.0	73	18.25	99	24.75	69		95		121	
23	1.97	49	10.64	49	12.25	74	18.5	100	25.0	70	20.4	96	28.0	122	35.6
24	5.2	50	10.85	50	12.5	75	18.75	101	25.25	71		97		123	
25	5.42	51	11.85	51	12.75	76	19.0	102	25.5	72	21.0	98	28.6	124	36.2
26	5.64	52	11.29	52	13.0	77	19.25	103	25.75	73		99		125	
27	5.85	53	11.5	53	13.25	78	19.5	104	26.0	74	21.6	100	29.2	126	36.8
28	6.07	54	11.72	54	13.5	79	19.75	105	26.25	75		101		127	
29	6.29	55	11.91			80	20.0	106	26.5	76	22.1	102	29.8	128	37.3

Key
Columns numbered 1 are total of duct dimensions width plus height or W + H = column no. 1 (see text).
Columns numbered 2 are the weight of metal required to make one linear foot of the indicated duct size (see text).
Additional weights must be added to the totals to account for waste from fabrication (see text).

Figure 27.1   Duct take-off factors

GALVANIZED AND HOT ROLLED MILD STEEL SHEETMETAL THICKNESS AND WEIGHTS				
U.S. Gauge	Thickness Inches		Weight Pounds per Sq. Ft.	
	Hot Rolled	Galvanized	Hot Rolled	Galvanized
24	.0239	.0276	1.00	1.5625
22	.0299	.0336	1.25	1.40625
20	.0359	.0396	1.50	1.65625
18	.0478	.0516	2.00	2.15625
16	.0598	.0635	2.50	2.65625
14	.0747	.0785	3.125	3.28125
12	.1046	.1084	4.375	4.53125
10	.1345	.1382	5.625	5.78125

Figure 27.2 Sheet metal gauges and thicknesses for galvanized and hot rolled mild steel sheets

The following example demonstrates the way to use the duct factor sheet. A duct 12 inches by 14 inches is described as 12″ x 14″. (The twelve inches refers to the side of the duct that faces the viewer.) The total of these dimensions is 12″ + 14″ = 26″. This 26″ is found in Column 1 on the chart (see Figure 27.1).

Since the larger dimension of the duct size is 14″, the duct should be fabricated from 24-gauge metal. Move from the figure in Column 1 (at the number 26) to the right, to Column 2. The number 5.64 is found in Column 2 beside the 26 in Column 1. This number (5.64) represents the pounds of metal required to make one linear foot of 12″ x 14″ duct-work using 24-gauge metal.

This same procedure is followed for each size of ductwork shown on the plans.

27.5 MEASURING THE LINEAR FEET OF DUCTWORK REQUIRED

On the ductwork chart (see Figure 27.1), information under Column 2 lists the pounds of metal required to fabricate one linear foot of ductwork of the indicated size. To find the number of linear feet of duct-work during the sheet metal take-off, special care should be taken in measuring linear feet of straight ductwork. Each partial foot less than six inches should be omitted. Each partial foot six inches or more should be counted as a full foot.

27.6 MEASURING LINEAR FEET FOR DUCT FITTINGS

When ductwork fittings are fabricated, considerably more sheet metal is wasted in the process than is normally wasted in fabricating straight ductwork. To offset the additional waste of sheet metal during fabrication of ductwork fittings, each fitting in the duct system should be measured from each direction. When this is done, the number of linear feet for the fitting is essentially doubled (see Figures 27.3–27.8).

When the duct dimension is changed in a fitting (such as a transition fitting), each duct size is used to determine the linear feet of ductwork for the fitting (Figure 27.3).

When the fitting is measured from each direction, the extra metal (waste) required for the fitting has been allowed for in the take-off. Examples of ways to measure elbow fittings are shown in Figures 27.3–27.8. The measuring procedure is the same for a square throat elbow and a round radius elbow. If the duct size is changed with the elbow, both duct sizes are used when showing the length of duct for the elbow (Figures 27.5 and 27.6). An offset is measured like a transition (Figures 27.4 and 27.7).

One rule is important enough to repeat: if the duct size changes in a fitting, one linear measurement across the fitting is made using one duct dimension and the second measurement across the fitting is made using the other duct dimension.

Figure 27.3 Measuring a splitter fitting for estimating ductwork costs

Figure 27.4 Measuring transition fitting for estimating ductwork costs

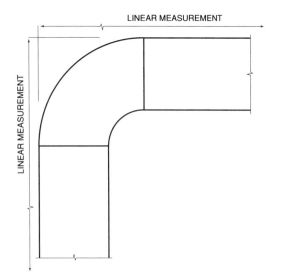

Figure 27.5 Measuring a round radius elbow for estimating ductwork costs

Figure 27.6 Measuring square elbow for estimating ductwork costs

The splitter fitting with two branches is measured somewhat differently from the other fittings described. Each duct size is measured as shown in Figure 27.8. The linear measurement of the straight ductwork is measured, and this measurement is extended in a straight line to where a new duct size is given. For example, the 12″ x 12″ straight duct is measured and the measurement is extended through the fitting to the 16″ x 12″ duct joint as shown (see Figure 27.8.) This is repeated for each duct size. Measuring the fitting four times adds enough metal for the fabrication and installation of this more complicated duct fitting.

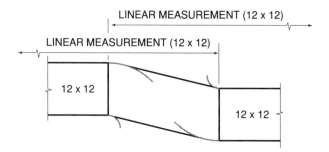

Figure 27.7 Measuring an offset fitting for estimating ductwork costs

Figure 27.8 Measuring a 3-Way splitter for estimating ductwork costs

27.7 DUCTWORK TAKE-OFF FORM

The take-off person must have a form that keeps up with the duct take-off and facilitates tabulation of the total weight of metal needed to fabricate the ductwork. The take-off sheet can be an original design or a variation of the take-off sheet shown in Figure 27.9. This sheet is used in the examples in this chapter.

Column A on the sheet indicates the duct dimensions. Column B provides space to total the two duct dimensions. Column C provides space for the measured length of the duct size indicated in Column A. This total measured length of duct uses the double measuring of the fittings described in Section 27.6 above. The gauge of metal required for the duct size is recorded in Column D. Column E provides space

to enter the weight of metal required to fabricate one foot of the duct size indicated. This weight is listed in Column 2 of Figure 27.1. Column F provides space for the total weight of the duct size indicated. Column F contains the poundage resulting from multiplying the total measured length (shown in Column C) times the weight of metal required per linear foot of duct (shown in Column E). Column G is used to indicate the total pounds of metal required for a specific length and size of ductwork. The formula (C x E = G) can be used to determine the total pounds of metal required to make the measured length of ductwork of the measured size.

27.8 DETERMINING THE TOTAL WEIGHT OF METAL REQUIRED FOR DUCTWORK

The procedure explained in Section 27.7 is repeated for each size of ductwork shown on the design plans. When all ductwork has been taken off and the multiplication of linear feet times weight per linear foot has been completed, the total weight of metal can be calculated. To obtain the total of metal weight required for the job, add together all numerals in Column G on all take-off sheets.

When the total weight for the metal has been determined, a percentage must be added to that total weight to allow for the overall waste in fabricating the ductwork and for the miscellaneous metal required for hangers and trim needed for hanging the ductwork on the job. (Remember that some extra metal has been figured in the fitting take-off to account for the additional metal needed for making the fittings versus straight ductwork.)

The extra percentage added for waste and miscellaneous metal must be added to the total weight of metal tabulated on the take-off sheet. The waste percentage depends on the degree of efficiency of the sheet metal shop, and the company makes this percentage figure available to the ductwork technician. Some companies figure a waste factor of approximately thirty percent. If special hangers or sheet metal specialties are needed to install the ductwork, these items must be taken off (measured) and priced separately from the ductwork.

A DUCT SIZE	B DUCT LENGTH PLUS WIDTH	C DUCT MEASURED LENGTH	D METAL GAUGE	E WEIGHT PER LINEAR FOOT	TOTAL WEIGHT
				TOTAL LBS. METAL	

Figure 27.9   Duct take-off form

An example of the metal take-off and the adding of the waste factor is as follows:

The metal take-off shows a total of 2,000 pounds of metal. With a 30% waste factor, the waste poundage is 2,000 x 30% (or 600 pounds of metal). When this 600 pounds is added to the original 2,000 pounds of metal, the grand total of metal needed for the job is 2,600 pounds.

27.9   COST FOR MATERIALS FOR SHOP-FABRICATED DUCTWORK

By multiplying the grand total poundage of metal times the cost of the metal per pound, the total cost of the metal required for the ductwork can be determined. For example, 2,600 pounds x $1.00 per pound = $2,600.00 for the cost of the metal.

27.10   LABOR COSTS FOR SHOP-FABRICATED DUCTWORK

The cost of fabrication and installation of the ductwork can be determined by multiplying the grand total poundage of metal times the labor cost factor for fabrication and installation. This labor cost factor should be available from the company. For example: 2,600 pounds of metal x $1.10 per pound labor cost = $2,860.00.

27.11 DUCTWORK INSULATION

Several different materials are used to insulate ductwork. The more popular materials are fiberglass, mineral wool, and cellular insulation. There are two basic types of insulation: exterior insulation (sometimes called wrapped insulation) and interior insulation that is glued inside the ductwork (called duct liner). Although this chapter is not concerned with the *materials* or *types* of insulation, a method for taking off and estimating the cost of insulation will be explained.

Insulation take-off procedures are the same regardless of the material or type of insulation. Determining the total square feet of insulation is the job of the take-off person. It is important to keep the figures for the different types and different materials separate because the cost varies for these different insulations. To calculate accurately, prices for each kind of insulation must be figured separately.

27.12 METHOD FOR TAKING OFF DUCTWORK INSULATION

The process for the linear take-off of insulation is identical to the ductwork take-off of sheet metal. The duct size, the lengths of straight duct, and the two-way take-off for fittings are the same as described for the sheet metal take-off (Figures 27.3–27.8).

27.13 READING THE INSULATION CHART

The duct insulation factors chart (Figure 27.10) shows Column 1 and Column 2 and their continuations. Column 1 represents the total duct dimensions (length plus width). A duct 18" by 12" totals 30 inches. Find the number 30 on Column 1 of the chart. To the right of this number, in Column 2, the number 5.0 is listed. This number indicates that five square feet of insulation are needed to cover one linear foot of ductwork eighteen inches by twelve inches in size. This procedure is repeated for all ductwork on the plan.

DUCT INSULATION FACTORS CHART

1	2	1	2	1	2	1	2	1	2
12	2.0	37	6.17	62	10.32	87	14.52	124	20.7
13	2.17	38	6.33	63	10.5	88	14.7	126	21.0
14	2.33	39	6.5	64	10.65	89	14.85	129	37.9
15	2.5	40	6.67	65	10.85	90	15.0	130	
16	2.66	41	6.85	66	11.0	91	15.2	131	38.5
17	2.83	42	7.0	67	11.15	92	15.35	132	
18	3.0	43	7.16	68	11.3	93	15.52	133	39.0
19	3.27	44	7.35	69	11.5	94	15.7	134	
20	3.34	45	7.5	70	11.65	95	15.85	135	39.6
21	3.5	46	7.68	71	11.84	96	16.0	136	
22	3.67	47	7.84	72	12.0	97	16.2	137	40.2
23	3.83	48	8.0	73	12.18	98	16.35	138	
24	4.0	49	8.16	74	12.33	99	16.5	139	40.8
25	4.17	50	8.32	75	12.5	100	16.7	140	
26	4.34	51	8.5	76	12.7	102	17.0	141	41.4
27	4.5	52	8.66	77	12.84	104	17.4	142	
28	4.67	53	8.84	78	13.0	106	17.7	143	42.0
29	4.83	54	9.0	79	13.2	108	18.0	144	
30	5.0	55	9.18	80	13.4	110	18.4	145	42.6
31	5.17	56	9.35	81	13.5	112	18.7	146	
32	5.35	57	9.5	82	13.7	114	19.0	147	43.2
33	5.5	58	9.65	83	13.85	116	19.4	148	
34	5.67	59	9.85	84	14.0	118	19.7	149	
35	5.84	60	10.0	85	14.2	120	20.0	150	
36	6.0	61	10.16	86	14.35	122	20.4	151	

KEY

Columns numbered 1 are total of duct dimensions width plus height or W + H = column no. 1 (see text).

Columns numbered 2 are the square footage of insulation required to insulate one linear foot of duct of the indicated size.

Additional square footage of insulation must be added to the grand totals to account for waste in fabrication (see text).

Figure 27.10 Duct insulation factors

27.14   INSULATION TAKE-OFF FORM

An insulation take-off form is needed to keep the take-off orderly and correct (Figure 27.11). This form is similar to the one described for sheet metal take-off. Column A is for the duct dimension (duct size). Column B provides space for the total of the duct dimensions (length plus width). Column C provides space for the measured length of duct. This measured length of duct contains the measured length of the fittings (measured in both directions). Refer to Figures 27.3–27.8, pages 182 and 183, for examples. The duct length should be measured as described for sheet metal take-off. Column D provides space to enter the square feet of insulation per linear foot of duct for the duct size listed (taken from the insulation chart, Figure 27.11). Column E

provides space to enter the total square feet of insulation needed for the indicated duct size.

When the measurements in Column C and Column D have been recorded on the take-off sheet, the take-off person multiplies linear feet (Column C) times square feet of insulation (Column D) to obtain the square footage of insulation required. This figure is entered in Column E. In other words, Column C x Column D = Column E.

After the multiplication step, the amounts in Column C are totaled to determine total square footage of different types of insulation required for the job. *Special care must be taken to separate different kinds of insulation in the take-off because the different insulations have different prices per square foot.*

A percentage is usually added to the total square footage of each type of insulation to account for waste that occurs during the installation process. This percentage varies from shop to shop; however,

A DUCT SIZE	B DUCT LENGTH PLUS WIDTH	C DUCT MEASURED LENGTH	D SQ. FT. PER LINEAR FOOT	E TOTAL SQUARE FEET

Figure 27.11   Insulation take-off form

25 to 30 percent is sometimes used as a reasonable amount to be figured for waste.

27.15 COST FOR INSULATION MATERIALS

To determine the total cost of insulation material, multiply the total square feet of insulation (including the waste factor) by the cost of the various types of insulation required. Sometimes it is necessary to add additional cost for special requirements such as plaster, special fasteners, and so forth. This cost is part of the insulation material cost.

The labor cost for installing the insulation is determined by the air conditioning and heating shop. Each shop establishes a cost figure per square foot of insulation installed.

27.16 COSTS OF LABOR FOR INSTALLING INSULATION

When the cost per square foot of insulation is multiplied by the total square feet of insulation, the labor cost for installing the insulation is obtained. Because labor costs vary for various types of insulation, different types of insulation should be calculated separately.

Add any special requirements to the labor cost. This cost must be calculated to suit job requirements and conditions.

SUMMARY

• The kind of ductwork is noted and described on the plans, in the specifications, or in both places.

• Sheet metal gauge is determined by its thickness.

• Estimation of size and cost of ductwork, fittings, and insulation is easier if the take-off person understands and uses appropriate charts.

• The amount of sheet metal needed must include an extra amount to cover the waste factor during fabrication, especially of fittings and transitions.

• Different types of ductwork and insulation materials must be figured separately because of price variations.

• The percentage of waste factor and labor cost for fabrication varies from shop to shop.

REVIEW QUESTIONS

1. List three materials used for fabricating ductwork.

2. Where is the kind of material for ductwork specified?

3. What does the term "take off" mean?

4. What is meant by the "gauge" of sheet metal?

5. Describe (or draw an explanatory sketch) how fittings and transitions are measured.

6. As used in this chapter, what does the formula C x E = G mean?

7. Why is extra material required for fittings and transitions?

8. Why should costs of different materials used for ductwork or insulation be figured separately?

9. What are the two basic types of insulation?

10. What are the five column headings of the ductwork take-off form used in this chapter?

11. What are the five column headings of the insulation take-off form used in this chapter?

12. Where does the take-off person get the labor cost for fabricating and installing ductwork and insulation?

ADDITIONAL STUDENT EXERCISES: Take-off Practice

1. Refer to Figure 27.12 to complete this exercise. Use the architect's scale and the ductwork take-off sheet to determine the total weight of metal required to fabricate the duct system shown. The duct system is drawn to $\frac{1}{4}$" = 1'-0".

2. Refer to Figure 27.12 to complete this exercise. Use the insulation take-off chart to estimate the square feet of insulation required for the duct system.

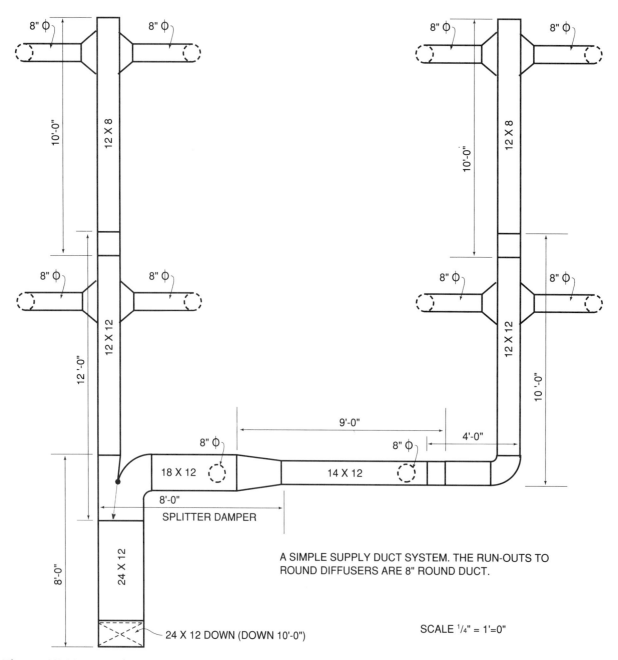

Figure 27.12   Exercise 1

28

READING MECHANICAL PLANS

OBJECTIVES

After studying this chapter, you should be able to:

- locate the title and scale used on the mechanical plan sheet

- scan a set of plans for an overview of the mechanical system

- locate heating and air-conditioning units, ductwork, and supply and return outlets

- identify and explain the Air Distribution Schedule.

- locate the mechanical specifications

- identify switches, thermostats, piping, and other components of the mechanical system

- answer specific questions about details of the mechanical system

NOTE: Refer to the plans for "A Small Bank and Trust Company" (at the end of this textbook) while studying this chapter.

28.1 INTRODUCTION

The first thing that a technician should do upon receiving a set of plans for a building is to review all the plans, scanning each one to see what drawings are available, where major information is shown, and to get a general overview of the building. A review of the specifications (see Chapter 13) will help the technician learn more about the building and what is expected of the builder.

The set of building plans located at the end of this textbook is typical for a small building. This chapter will refer to these plans to demonstrate arrangement and location of various portions of information. Architectural plans for the same building were discussed in Chapter 26. In Chapter 28 the discussion will be directed to the reading of the mechanical plan. Sheet M-1 shows the mechanical floor plan for the building.

Considerable attention will be devoted to ductwork because it is such a vital part of a building project. The student is encouraged to review Chapter 27 for a method for "taking off" ductwork and ductwork insulation for estimating purposes.

Student exercises located at the end of this chapter are based on information shown on Mechanical Sheets M-1 and M-2.

28.2 READING THE MECHANICAL PLANS

The first thing to find on the sheet is the title of the drawing. On Sheet M-1 the title is "Floor Plan." If the building had multiple floors, the title would indicate which floor is represented on the drawing. The scale used for the drawings is given as part of the title. In this case, the plan has been reduced, making the $1/4"=1'-0"$ scale incorrect. Because of this discrepancy, the only dimensions that can be relied on are the ones specified on the plans. In actual practice, this drawing would not be issued as a construction drawing for builders and technicians to work from.

After determining the title of the plan and the scale used for the drawing, the technician should review the entire plan to study all the details and construction information that is given on the sheet. If there are multi-floor and multi-sheet sets of plans, it is important to know, in general, where various details, schedules, sections and so forth are located. It is also important to obtain an understanding of the entire mechanical system.

In reviewing this plan, it is suggested that the technician read each sheet like reading a book. This procedure is described in Chapter 26.

28.3   LOOKING FOR SPECIFIC SYSTEMS

After the general overview of the plan and the mechanical design, each specific part of the air-conditioning and heating system should be reviewed. In most cases, the air conditioning and heating units are reviewed first. On this building, the units are a part of a split system, air-to-air heat pump. The outside unit is located on the roof, and the inside unit is located in a mechanical space above the ceiling to the right of the front entrance. A floor plan of this mechanical space is shown in the upper left-hand corner of the mechanical plan. (See Sheet A-5, Drawing Section 10/2.A-1, A-2, and A-3 for location of mechanical space above the ceiling.)

28.4   FOLLOWING THE DUCTWORK

The inside section of the heat pump has a 30" x 18" supply duct extended toward the top of the plan. (Note: Where the North arrow is not indicated, the top of the sheet is normally considered north.) This supply duct splits into two 15" x 18" ducts with a splitter damper, and these two ducts continue to serve the air supply to the entire building. The supply air outlets are shown with the air quantities (CFM) indicated.

The return air duct connects to the inside section of the heat pump and extends to the front entrance, turns down and connects to an 18" x 18" return air grille mounted in the wall six inches below the ceiling.

The duct sizes are shown on the plans. In Chapter 27 it was pointed out that the first dimension shown on ductwork is the side of the ductwork that is being seen, and the other dimension is the second number. In other words, the 30" x 18" duct shown for the supply ductwork is 30 inches wide and 18

28.5   THE SUPPLY AND RETURN AIR OUTLETS

In the table at the top center of the mechanical plan is the Air Distribution Schedule. This schedule gives the description of the supply and return outlets including a typical manufacturer and catalog number. In many jobs, the design engineer or technician is not allowed to mention specific manufacturers for the specified equipment. In such cases, a detailed description of the desired air distribution outlet must be included in the specification.

28.6   WRITTEN NOTES AND INSTRUCTIONS

It is sometimes advantageous to have written notes and instructions placed on the plans. On this job these notes have been titled "General Notes."

28.7   OTHER INFORMATION SHOWN ON THE MECHANICAL PLAN

In addition to the information described above, the mechanical plan describes and shows electric heaters, timed override switches, night set-back thermostats, and refrigeration piping. Also included on the mechanical plan are such items as drawings of equipment space and sections. Plan 2/M-1, on sheet M-1 is a plan of the equipment space located over the Women's Toilet. Section 4/M-1 is a longitudinal section of the inside heat pump unit.

28.8 MECHANICAL SPECIFICATIONS

As noted in Chapter 13, some small jobs have the specifications printed on a separate sheet and bound with the regular plans. For example, the specifications for the plans at the end of this book are bound in this way. On larger jobs, all specifications are bound in a separate book that accompanies the plans when they are turned over to the contractor.

SUMMARY

• Information regarding the scale of the drawing and specific information on the sheet is found on the title of the sheet.

• It is important for the technician to gain an understanding of the entire mechanical system before concentrating on any of its parts.

• The types of air conditioning and heating units and their locations are shown on the mechanical plans.

• If there is no directional arrow on the plans, the top of the plan is considered north.

• The technician should study the supply and return air ductwork throughout the system.

• When ductwork sizes are stated, the first number refers to the visible side of the ductwork, and the second number refers to the other dimension.

• The Air Distribution Schedule describes supply and return outlets for typical units or includes a detailed description in the specification.

• Written notes and instructions may be placed on the plans.

REVIEW QUESTIONS

1. Upon receiving a set of plans, what information should the technician look for first?

2. What information is found on the title of the drawing?

3. Why is the scale listed on the drawings at the end of the textbook not accurate?

4. What kind of heating and air-conditioning units are used in the small bank and trust company?

5. Where are the units located?

6. Where is the mechanical space shown on this mechanical plan?

7. If there is no directional arrow on the plan, which direction is considered north?

8. What are the dimensions of the two supply ducts that serve the air supply to the building?

9. What are the dimensions of the main return air ducts?

10. On a duct with dimensions of 30" x 18", what does the 30" refer to?

11. What information is given in the Air Distribution Schedule?

12. If the designer is not allowed to mention specific manufacturers, how are descriptions of outlets given?

13. What other types of mechanical equipment are shown on the mechanical plan?

14. On small jobs, where may mechanical specifications be placed?

15. Where are the mechanical specifications usually placed on large jobs?

ADDITIONAL STUDENT EXERCISES: Reading Mechanical Plans

A. Refer to Sheet M-1 in the back of the book to answer the following questions.

1. How many exhaust fans are located on the roof?

2. What is the cooling capacity of the heat pump?

3. What rooms have electric heaters installed?

4. Where are the refrigeration lines shown to be installed?

5. What are the sizes for the refrigeration lines?

6. How many thermostats are needed for the job? Name the types of thermostats needed.

7. What is the symbol for the supply outlet in the conference room? What is the symbol for the supply outlet for the East entrance?

8. What is the total air (CFM) that the system is designed to distribute throughout the building?

9. Are dampers installed in the duct take-offs for balancing the air quantities?

10. What is the minimum clearance around the outside heat pump unit?

11. A time clock is shown as part of the control system. Where is this clock located?

12. Does the time clock have an override switch? If so, where is this switch mounted?

13. What is the size of the condensate drain line?

14. Are roof curbs fabricated on the job?

15. Is the outside heat pump unit located on the roof?

16. Does the exhaust fan have a damper?

17. Where are the starter and disconnect switches for the inside heat pump unit located?

18. Are any square or rectangular duct offsets required?

19. How high (above the floor) should the thermostat be mounted?

B. Refer to Sheets M-1 and M-2 of the plans at the end of the textbook to answer the following questions.

1. How many fresh air intakes are shown?

2. What size flexible ductwork is used?

3. What is the size of the largest duct shown?

4. What air distribution and return air grilles are specified?

5. Is there a disclaimer stated regarding the plans?

6. List the codes, agencies, and so forth that the contract work must comply with.

7. What are special requirements for the exhaust fan?

8. Is the inside heat pump unit floor-mounted?

9. Describe the unit casing.

10. Is the air-handling unit insulated?

11. Describe the supply air fan.

12. What special features should the compressor have?

13. What special fittings are to be installed in the equipment lines off the compressor?

14. What type piping is used for the condensate drain line?

15. What type and thickness insulation is used on the refrigerator lines?

16. What agency is specified for standards for constructing the ductwork?

17. What is the maximum spacing for the horizontal duct hangers (supports)?

18. Where are flexible duct connectors used?

19. What is the time clock used for?

20. Does the time clock also control the exhaust fan ON-OFF schedule?

21. Who does the testing of the equipment upon completion of the job?

22. Is a warranty required? If so, describe it.

29

READING ELECTRICAL PLANS

OBJECTIVES

After studying this chapter, you should be able to:

- explain how the various types of lighting fixtures are identified on the electrical plan
- discuss the information about electrical wiring as shown on the electrical plan
- identify and explain symbols, notes, and schedules on the electrical plans
- define electrical specifications
- discuss reasons why coordination by the various tradespersons is necessary

NOTE: Refer to the plans for "A Small Bank and Trust Company" (at the end of this textbook) while studying this chapter.

29.1 INTRODUCTION

As noted in Chapter 28, the first thing that a technician should do upon receiving a set of plans for a building is to review them, scanning each one to get an overview of the building. The electrical system, the air-conditioning and heating system, and the plumbing system for a building are very closely related. Design technicians and installation technicians must coordinate the systems of all trade areas to have a well-designed and smooth installation on the job. Chapter 26 discussed the architectural plans that are at the end of this textbook. Chapter 28 concerned the reading of the

mechanical plans. This chapter will deal with the reading of the electrical plans.

On larger buildings, the electrical plans usually contain plans for lighting, power distribution (showing receptacles and special connections), panel schedules, and other schedules and details pertaining to the electrical system. Since the construction plans at the end of this book are for a small building, all the necessary information is shown on one sheet.

29.2 READING THE ELECTRICAL PLANS

The larger drawing on Sheet E-1 is the floor plan. The lighting plan and the power distribution plan are consolidated on the same floor plan. Lighting fixtures are shown with rectangles for fluorescent fixtures and circles for other types of fixtures. Each lighting fixture is identified by a letter of the alphabet. These letters are shown on the lighting fixture schedule on the plan.

The power distribution is also shown on the plan. The electrical receptacles use the standard symbol (see Chapter 7 and Appendix D). The hash marks on the circuit lines indicate the number of conductors (wires) required. All of the circuits are indexed (numbered), and these numbers indicate the number of the circuit breaker in the distribution panel. A schedule for Panel A is shown on the plan with each circuit numbered. The service to the panel is shown with a separate detail. The main circuit breaker is located inside Panel A.

29.3 THE ELECTRICAL WIRING

The electrical wiring is shown with an arched line to electrical devices and arched lines between lighting fixtures or between receptacles. The exact location of the wiring and conduit is left to the installing technician. On each wire are hash marks indicating how many wires are required with the circuit. The arrowhead at the end of the wiring symbol indicates that the conduit is to be extended to the panel. The small number located at the arrowhead gives the number of the circuit breaker to which the wiring is to be connected. The panel detail located in the lower left-hand side indicates the circuit breaker numbers. The wiring shown with broken lines, and the receptacles connected with these broken lines, are alternate (marked ALT). The ALT marking means that the contractor should give a separate price for this work, and the owner can decide whether to include this work in the contract.

29.4 ADDITIONAL INFORMATION SHOWN ON ELECTRICAL PLAN

The electrical plan on Sheet E-1 contains additional information that should be noted at this point. A lighting fixture schedule on the plan describes the lighting fixtures that are to be used on the job. Each fixture shown on the floor plan has a letter that corresponds to the fixture schedule.

A symbol schedule identifies the symbols used on the plan. The power riser diagram is shown in the lower right-hand corner. A general note is also shown on the plan giving instructions to the contractor. A note at the top right corner of the plan restricts the use of the plans.

29.5 ELECTRICAL SPECIFICATIONS

The electrical specifications, like the mechanical specifications, contain written instructions and descriptions about construction equipment and procedures. The specifications for this job were bound with the General Contract specifications and are not available with this text.

29.6 UNDERSTANDING THE PLANS

As previously noted, it is very important for persons reading the plans to understand the plans thoroughly before attempting to estimate the cost of the job or to construct the job. After construction is begun, the various trades often have to do their installations in the same limited space. For this reason, coordination on the job between the different contractors is necessary to complete the job satisfactorily.

SUMMARY

• Coordination among the various tradespersons is necessary for a successful construction project.

• Lighting fixtures are identified by symbols and letters on the lighting schedule.

• Electrical receptacles, circuits, and circuit breakers are shown on the power distribution plan.

• Specific information regarding the electrical wiring is given on the plan, but the exact location of wiring and conduit is usually left up to the installing technician.

• The exact description of each lighting fixture is given on the fixture schedule.

• Electrical specifications give instructions about equipment and procedures and are sometimes listed under the General Contract specifications.

• It is important to understand plans thoroughly before attempting to estimate a job or construct a project.

REVIEW QUESTIONS

1. Why should the technician scan the entire set of plans before beginning any work?

2. What kind of lighting does a rectangle indicate? A circle?

3. Draw the standard symbol for an electrical receptacle.

4. Where is the main circuit breaker located?

5. What do the hash marks on the circuit lines represent?

6. Who decides the exact location of wiring and conduit?

7. What does ALT mean? When would ALT be found on the plans?

8. What information is given on the note at the top right corner of the plans on Sheet E-1?

9. Why are no electrical specifications shown on the plans on Sheet E-1?

ADDITIONAL STUDENT EXERCISE: Reading Electrical Plans

A. Refer to Sheet E-1 in the back of the book to answer the following questions.

1. Is any electric baseboard heating to be installed on this job?

2. Is there an exhaust fan on the job?

3. Is the electrical wiring shown on the floor plan?

4. Does the job have a telephone space board?

5. Does the outside section of the heat pump have a disconnect switch?

6. How is the outside lighting controlled?

7. How many telephone outlets are shown on the electrical plans?

8. How many above-the-counter receptacles are shown?

9. Is there a separate space for the amplifier and tuner for the drive-up window?

10. Are any electrical receptacles located in the floor?

11. Why are some of the electrical receptacles (marked ALT) and the "home runs" shown with broken lines?

12. Why is the soffit lighting (outside) put on a time switch?

13. Where is the telephone space (wiring board) located?

14. Does the mechanical contractor do the control wiring for the HVAC equipment?

15. How many lighting tubes are to be installed in each Type A fixture?

16. Is there a duct-mounted electric heater on the job?

B. Refer to Sheet E-1 of the plans at the end of this textbook to answer the following questions.

1. How many Type A lighting fixtures are shown?

2. How many circuits are being used for lighting?

3. What is the connected load for the baseboard heater?

4. What electrical service is connected to Electrical Panel "A"?

5. What electrical load is connected to the duct heater?

6. What circuits are connected to the outside section of the heat pump?

7. What size wires are used to connect the electrical service to the circuit breaker for the outside section of the heat pump?

8. Is the electrical wiring that serves the outside section of the heat pump installed in conduit?

9. Who makes final electrical connections to the mechanical equipment?

10. What is the maximum number of circuit breakers that can be put on the electrical panel?

11. How many Type F fixtures are shown?

12. Where are electrical receptacles mounted flat in the corners?

13. Where is the volume control for the speaker system in the lobby located?

14. The second volume control is located in the office near the storage room. How many speakers does it serve?

30

READING PLUMBING PLANS

OBJECTIVES

After studying this chapter, you should be able to:

- explain why coordination by the various trades is necessary
- list the various plumbing systems found on the plumbing plans
- identify the various types of piping on the plans
- explain what information is found on the Plumbing Fixture Schedule
- discuss the various line configurations used to show specific types of piping
- identify the parts of the plumbing riser diagram

NOTE: Refer to the plans for "A Small Bank and Trust Company"(at the end of this textbook) while studying this chapter.

30.1 INTRODUCTION

As noted in Chapters 28 and 29, the first thing that a technician should do upon receiving a set of plans for a building is to review them, scanning each plan to get an overview of the building. Understanding the mechanical, electrical, and plumbing plans is very important because the three trades will be working in the same area to install their equipment, and coordination of their work will be required. Studying each of the plumbing systems one at a time will help the technician understand the entire system more thoroughly.

The major systems found in a complete plumbing plan are as follows: sanitary drainage system, plumbing vent system, domestic hot and cold water system, roof drainage system, fire protection sprinkler system, and compressed air system. In a hospital or medical building, plumbing includes the medical gas systems (oxygen, medical air, nitrogen and vacuum). All of these systems will occupy the same space that is allotted for the mechanical and electrical systems. It is essential, therefore, that the systems be understood and coordinated. The set of plans at the end of this book is typical for a small building, with only one sheet required for the plumbing system. This chapter will focus on the plumbing plan for this building.

30.2 LOOKING FOR SPECIFIC SYSTEMS

On the plan, the size and location of each piping system is shown. The domestic cold water piping includes the service piping to the building and the distribution piping inside the building. The water heater and the domestic hot and cold water piping are shown.

The heavier lines represent the drain lines. The lines connected to the plumbing fixtures are the sanitary waste lines. The sanitary piping is located underground except for the part that turns up to connect to the plumbing fixtures.

The roof drainage piping is connected to the roof drains. This piping is extended, usually above the ceiling, to a point where it can be turned down and installed below the floor. These lines are drawn the same as the sanitary drain lines, and they can be identified as storm drain lines because they are connected to the roof drains.

Another piping system is the plumbing vent piping. This piping is connected above the trap on each fixture to let air enter and leave the piping system when water is introduced into the drainage system. (Remember that water and air cannot occupy the same space.)

30.3 READING THE PLUMBING PLANS

The floor plan shows the location of the plumbing fixtures, and the fixtures are numbered to correspond to the numbers in the Plumbing Fixture Schedule. The schedule has the name of the fixtures, the manufacturer and model number of each fixture, and the connection size for all the piping connected to the fixture. There is a space for notes that pertain to the fixtures.

The domestic cold water piping is shown with a light solid line having long dashes and a dot. The domestic hot water is shown with the same weight line as the domestic cold water lines except that the line is a series of long dashes with two dots. These piping systems have all the fittings and valves needed for a complete system.

The sanitary drainage piping is shown as a heavy solid line. All the various fittings and related equipment associated with the drainage piping are shown. The plumbing vent piping is shown with a series of dashes. The weight of these dashes is the same as the weight for the sanitary drain line.

The roof drainage system consists of a series of roof drains with related piping and fittings. This piping is shown with the same weight lines as the sanitary drain lines.

Notes are used to direct the technician during the installation. Because this is a small job, the specifications are shown on the drawing. These specifications are titled "Plumbing Notes and Specifications."

A plumbing riser diagram on the plan shows the method for installing the various piping systems serving the plumbing fixtures. The pipe sizes are shown on this diagram along with the method for connecting them. This diagram has the sanitary piping, vent piping, and domestic hot and cold water. Each vent riser is identified with an "R" number or riser number.

The connection to the water heater is included. A larger connection detail for the water heater is shown in the upper left-hand corner of the drawing. This detail shows the piping and fittings necessary for connecting the piping to the heater.

Two other details are shown on the plan. The installation of the exterior clean-out is shown with notes about the installation. The other detail shows the method for installing the emergency roof drain. The purpose of this drain is to drain water off the roof if the other drains fail, or if for some reason water builds up on the roof to a depth of three inches.

SUMMARY

• Understanding mechanical, electrical, and plumbing plans is necessary for each tradesperson.

• The plumbing plan includes sanitary drainage, plumbing vents, domestic hot and cold water, roof drainage, fire protection sprinkler, and compressed air systems.

• In a hospital or medical facility, the medical gases (oxygen, medical air, nitrogen, and vacuum) are part of the plumbing.

• The plumbing plan shows whether piping is underground or above ground and includes the connections and location of piping.

• The Plumbing Fixture Schedule identifies the manufacturer, model number, and other pertinent information about each fixture.

• Certain line weights and configurations denote various types of piping.

• The riser diagram shows how various piping systems are to be installed.

• Details show specific information on connections, installation, and fittings.

REVIEW QUESTIONS

1. Why does the plumbing technician need to be familiar with the mechanical and electrical plans?

2. List six parts of the complete plumbing system.

3. List four types of medical gases found in hospitals and medical facilities.

4. List a significant fact about each of the following:
   A. Domestic cold water piping
   B. Sanitary waste lines
   C. Roof drainage piping
   D. Plumbing vent piping

5. Draw the distinctive lines that indicate each of the following:
   A. Domestic cold water piping
   B. Domestic hot water piping
   C. Sanitary drain piping
   D. Plumbing vent piping

6. On the "Small Bank and Trust Company" plans, where are the specifications located?

7. What does an "R" number indicate?

8. What information is found on the Plumbing Fixture Schedule?

9. What is the purpose of the emergency roof drain?

ADDITIONAL STUDENT EXERCISE: Reading Plumbing Plans

Refer to Sheet P-1 in the back of the book to answer the following questions.

1. How many water closets are required for the job?

2. How many lavatories are required for the job?

3. How many roof drains are shown?

4. How many exterior clean-outs are shown on the plan?

5. Are the domestic hot and cold water lines that are located inside the building insulated? If so, how thick should fiberglass insulation be?

6. Are all the roof drains the same?

7. Are the roof leaders insulated?

8. What is the size of the plumbing vent through the roof?

9. How many yard hydrants are required for the job?

10. Does the water heater have a safety relief valve?

11. What is the size of the cold water service?

12. What is the size of the sanitary sewers?

13. Does the cold water service have a shut-off valve?

14. Are roof leaders to be installed without regard to ductwork?

15. How many floor drains are shown?

16. Who wires the water heater?

17. What kind of piping is used for the roof drain piping?

18. Are domestic water lines sterilized?

19. Do hot and cold water service lines to the plumbing fixtures have air chambers?

20. Describe the emergency roof drain.

21. Where is the emergency roof drain located?

22. Do outside clean-outs have concrete pads?

APPENDICES

APPENDIX A

GEOMETRIC FIGURES AND FORMULAS

RECTANGLE

$A = L \times W$
$P = 2L + 2W$

SQUARE

$A = L \times W$
$P = 2L + 2W$

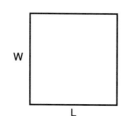

PARALLELOGRAM

$A = H \times L$

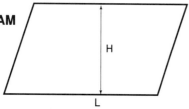

TRAPEZOID

$A = H \times \dfrac{L_1 + L_2}{2}$

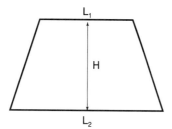

TRIANGLE

$A = \dfrac{L + H}{2}$

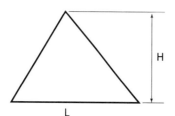

CIRCLE

$A = 3.14 \times R \times R$
$C = 3.14 \times D$
$R = \dfrac{D}{2}$
$D = R + R$

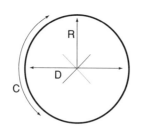

RECTANGULAR TANK

$A = 2 (W \times H) + 4 (L \times H)$
(TOTAL OUTSIDE AREA)
$V = W \times H \times L$

CYLINDER

$S = 6.283 \times R \times H + 6.283 \times R \times R$
(SURFACE AREA OF TANK)
$V = 3.14 \times R \times R \times H$

ELLIPTICAL TANK

$V = 3.14 \times Z \times Y \times H$
$S = 6.283 \times \dfrac{\sqrt{Y^2 + Z^2}}{2} \times H + 6.283 \times Y \times Z$

(S = OUTSIDE AREA)

TRIANGULAR STRUCTURE

$\dfrac{V = W \times H \times L}{2}$

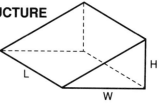

CONE

$S = 3.14 \times R \times S + 3.14 \times R \times R$
$V = 1.047 \times R \times R \times H$

KEY

S = SURFACE	C = CIRCUMFERENCE
A = AREA	P = PERIMETER
V = VOLUME	L & W = LENGTHS OF SIDES
Y = SHORT RADIUS	H = HEIGHT
Z = LONG RADIUS	R = RADIUS
S_H = SLOPE	D = DIAMETER

APPENDIX B

CONVERSION TABLES

Table 1. METRIC SYSTEM TO METRIC SYSTEM

10 millimeters	=	1 centimeter
10 centimeters	=	1 decimeter
10 decimeters	=	1 meter
10 meters	=	1 dekameter
100 dekameters	=	1 kilometer

Table 2. METRIC SYSTEM TO ENGLISH SYSTEM

1 millimeter	=	0.03937 inch
1 centimeter	=	0.3937 inch
1 meter	=	39.37 inches
1 kilometer	=	0.6214 miles

Table 3. ENGLISH SYSTEM TO METRIC SYSTEM

1 inch	=	25.4 millimeters	=	0.0254 meters
1 foot	=	304.8 millimeters	=	0.3048 meters
1 yard	=	914.4 millimeters	=	0.9144 meters
1 mile	=	1.6093 kilometers	=	1609.3 meters

Table 4. ENGLISH SYSTEM - CONVERTING FRACTIONS TO DECIMALS

INCH-Fractions		INCH-Decimals
$1/16$	=	0.0625
$1/8$	=	0.1250
$3/16$	=	0.1875
$1/4$	=	0.2500
$5/16$	=	0.3125
$3/8$	=	0.3750

Table 4. ENGLISH SYSTEM - CONVERTING FRACTIONS TO DECIMALS (CONTINUED)

INCH-Fractions		INCH-Decimals
$7/16$	=	0.4375
$1/2$	=	0.5000
$9/16$	=	0.5625
$5/8$	=	0.6250
$11/16$	=	0.6875
$3/4$	=	0.7500
$13/16$	=	0.8125
$7/8$	=	0.8750
$15/16$	=	0.9375

Table 5. CONVERSION TABLE - INCHES TO FEET

INCH		FOOT
1	=	.083
2	=	.17
3	=	.25
4	=	.33
5	=	.42
6	=	.50
7	=	.58
8	=	.67
9	=	.75
10	=	.83
11	=	.92
12	=	1.00

APPENDIX C

ABBREVIATIONS

ABBREVIATION	DESCRIPTION
A	Area
AC	Alternating Current
AD	Access door, Area drain
AP	Access panel
ADD	Addendum
AGA	American Gas Association
AIEE	American Institute of Electrical Engineers
AIR COND	Air Conditioning
AL	Aluminum
ALM	Alarm
AMP	Ampere
APPD	Approved
APPROX	Approximate
ARCH	Architect, Architectural
ASB	Asbestos
ASHRAE	American Society of Heating, Refrigeration and Air Conditioning Engineers
ASME	American Society of Mechanical Engineers
ASPH	Asphalt
ASRE	American Society of Refrigeration Engineers
ASSOC	Associate, Association
AVE	Average
AWG	American wire gauge
B	Bath, Brine

ABBREVIATION	DESCRIPTION
BB	Baseboard
BD	Board
BET	Between
BF	Back feed
BH	Boiler house
BK SH	Book shelves
BKR	Breaker
BLD, BLDG	Building
BLD LIN	Building line
BLO	Blower
BLR	Boiler
BO	Blow-off
BOT	Bottom
BP	Bypass
BR	Bedroom, Boiler room
BRK	Brick
BS	Bureau of Standards
BSMT	Basement
BT	Bathtub
BTR	Better
BTU	British thermal unit
BVL	Beveled
BWV	Back water valve
C	Thermal conductance
CB	Catch basin
CEM	Cement
CEM FL	Cement floor
CEM MORT	Cement mortar

ABBREVIATION	DESCRIPTION	ABBREVIATION	DESCRIPTION
CEM PLAS	Cement plaster	DIM	Dimension
CFLG	Counter flashing	DISC	Disconnect
CFM	Cubic Feet per minute	DMPR	Damper
CI	Cast iron	DO	Ditto
CIR	Circle	DR	Drain
CIR BKR	Circuit breaker	DW	Dry well
CKT	Circuit	DWG	Drawing
CL	Center line, Closet	DWN, DN	Down
CLG	Ceiling	E	East
CLR	Clear	EDR	Equivalent direct radiation
CND	Conduit	ELEV	Elevator
CO	Clean-out	EME	Emergency
CONC	Concrete	ENCL	Enclosure
COND	Conductor	ENGR	Engineer
CONN	Connection	ENT	Entrance
CONT	Contact	EQUIP	Equipment
CONTR	Contractor	EST	Estimate
CONV	Convector	EXIST	Existing
CONV ENCL	Convector Enclosure	EXT	Exterior
COP	Copper	F DR	Fire door
COV	Cover	F EXT	Fire Extinguisher
COV PL	Cover plate	FAB	Fabricate
CP	Cesspool	FBRK	Firebrick
CPM	Cycles per minute	FD	Floor drain
CR PL	Chrome plated	FDR	Feeder
CSG	Casing	FDT	Foundation
CTR	Center	FHY	Fire hydrant
CU FT	Cubic foot	FIG	Figure
CU IN	Cubic inch	FIN	Finish
CUR	Current	FIN FL	Finished Floor
CV	Check Valve	FL	Floor
CW	Cold water	FLG	Flange
CWP	Circulating water pump	FPM	Feet per minute
CYL	Cylinder	FPRF	Fireproof
D R	Dining room	FPS	Feet per second
D S, DS	Downspout	FR	Front
DC	Direct current	FT, '	Feet, foot
DEG, °	Degree	FTG	Footing
DF	Drinking fountain	FURN	Furnish
DIA, DIAM	Diameter	GA	Gauge
DIAG	Diagram	GALV	Galvanized
DIF	Diffuser	GL	Glass

ABBREVIATION	DESCRIPTION	ABBREVIATION	DESCRIPTION
GPH	Gallons per hour	MTG	Mounting
GPM	Gallons per minute	N	North
GR	Grade	NEC	*National Electric Code*®
GT	Grease trap	NFPA	National Fire Protection Association
GTV	Gate valve		
HB	Hose bibb	OFF	Office
HP	Horsepower	OVLD	Overload
HWH	Hot water heater	PAN, PAL	Panel
ID	Inside diameter	PC	Pull Chain
IN, "	Inch	PH	Phase, Power house
INFO	Information	PLUMB	Plumbing
INS	Insulate, insulation	PORT	Portable
INT	Interior	PREFAB	Prefabricated
INV	Invert	PROP	Proposal
J	Junction box	PRV	Pressure reducing valve
JCL	Janitor's closet	PS	Pipe shaft
K	Kitchen	PTD	Painted
KS	Kitchen sink	PWR	Power
KW	Kilowatt	R	Radius
KWHR	Kilowatt hour	RAD	Radiator
L	Left	RD	Road, Roof drain
LAD	Ladder	RECIR, RECIRC	Recirculate
LAU	Laundry	RED	Reducer
LAV	Lavoratory	REF	Refrigerator
LB, #	Pound	REG	Register
LB/CU/FT	Pounds per cubic foot	REM	Remove
LDG	Landing	REP	Repair
LEV	Level	RET	Return
LH	Left-hand	REV	Revision
LIB	Library	RF	Roof
LIN FT	Linear feet	RH	Right-hand
LKR R	Locker room	RM	Room
LR	Living room	RT	Right
LV	Louver	S	South
LWC	Light weight concrete	SEW	Sewer
M	Meter, motor	SH	Shower
MEZZ	Mezzanine	SL	Slate
MIN, '	Minute	SLD	Solder
MLDG	Moulding	SOV	Shut-off valve
MN	Main	SP	Static pressure
MRR	Men's rest room	SPKR, SPR	Sprinkler
MT	Men's toilet	SQ	Square

ABBREVIATION	DESCRIPTION	ABBREVIATION	DESCRIPTION
SQ FT	Square foot	VAC	Vacuum
SQ IN	Square inch	VB	Valve box
SS	Service sink	VEL	Velocity
SST	Stainless steel	VERT	Vertical
STD	Standard	VEST	Vestibule
SUP	Supply	VOL	Volume, volt
SV	Safety valve	VS	Vent shaft
SW	Switch	W	West, watt, wire
SWBD	Switchboard	WC	Water closet
SYS	System	WH	Hot water heater
T	Toilet	WH	Wall hydrant, weather head
TEMP	Temperature	WP	Waterproof
THERM	Thermometer	WRR	Women's rest room
TYP	Typical	WT	Weight, Women's toilet
UH	Unit heater	X SECT	Cross section
UL	Underwriters Laboratories	YD	Yard
UNFIN	Unfinished	YH	Yard hydrant
USSG	U. S. Standard Gauge	YDI	Yard drain inlet
V	Vent, volt	YR	Year

APPENDIX D

SYMBOLS

ARCHITECTURAL SYMBOLS

NOTE: The following are architectural symbols that are frequently used on construction drawings. Note that the architectural symbols for the same material are often different in the elevation view, plan view, and section view.

BUILDING MATERIALS			
	ELEVATION	**PLAN**	**SECTION**
EARTH			
BRICK			
CONCRETE			
CONCRETE BLOCK			
WOOD	SIDING STUD	STUD STUD	OR
GLASS	OR		
GLASS BLOCK			
INSULATION	OR		

213

ARCHITECTURAL SYMBOLS

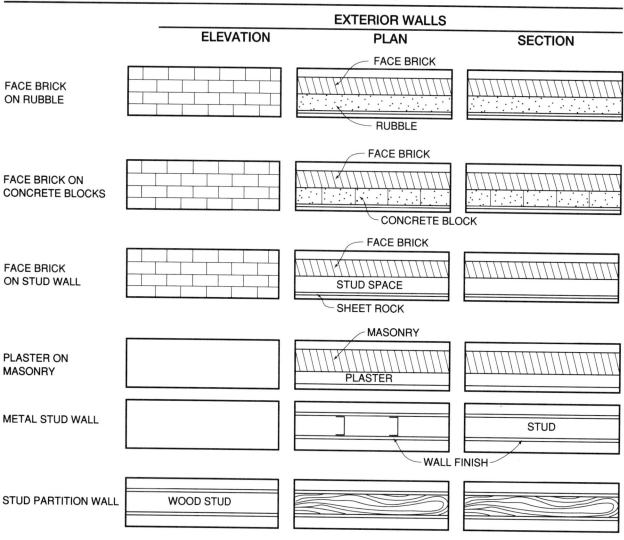

EXTERIOR WALLS

	ELEVATION	PLAN	SECTION
FACE BRICK ON RUBBLE		FACE BRICK / RUBBLE	
FACE BRICK ON CONCRETE BLOCKS		FACE BRICK / CONCRETE BLOCK	
FACE BRICK ON STUD WALL		FACE BRICK / STUD SPACE / SHEET ROCK	
PLASTER ON MASONRY		MASONRY / PLASTER	
METAL STUD WALL		WALL FINISH	STUD
STUD PARTITION WALL	WOOD STUD		

SITE PLAN SYMBOLS

DESCRIPTION	SYMBOL	DESCRIPTION	SYMBOL
NORTH ARROW	→N	WALK	
GRADE POINT		PROPERTY LINE	—— – ——
		FUEL GAS LINE	—— G ——
		CONTOUR LINE	
FIRE HYDRANT		POWER POLE WITH GUY	●———○
MANHOLE	MH	ELECTRICAL SERVICE LINE	—— E ——
TREE	+	WATER LINE	—— W ——
		FIRE LINE	—— F ——

SITE PLAN SYMBOLS (CONTINUED)

DESCRIPTION	SYMBOL
NATURAL GRADE (CONTOUR LINE) NUMBER INDICATES ELEVATION IN FEET ABOVE SEA LEVEL	*60*
FINISHED GRADE LINE	*60*

MECHANICAL SYMBOLS

DESCRIPTION	SYMBOL
RECTANGULAR OR SQUARE SUPPLY DUCT TURNED UP	
RECTANGULAR OR SQUARE SUPPLY DUCT TURNED DOWN	
RECTANGULAR OR SQUARE EXHAUST DUCT TURNED UP	
RECTANGUALR OR SQUARE EXHAUST DUCT TURNED DOWN	
ROUND SUPPLY DUCT TURNED UP	
ROUND SUPPLY DUCT TURNED DOWN	
RECTANGULAR CEILING DIFFUSER	
ROUND CEILING DIFFUSER	
FLEXIBLE CONNECTION	

MECHANICAL SYMBOLS (CONTINUED)

DESCRIPTION	SYMBOL
TRANSITION: FOT = FLAT ON TOP	FOT
ELECTRIC OPERATED DAMPER	E.O.D.
FIRE DAMPER	F.D.
SMOKE DAMPER	S.D.
ELECTRIC HEATER IN DUCT	
SUPPLY OUTLET WITH SIZE AND AIR QUANTITY SHOWN	12 X 8 / 200 CFM
DEFLECTOR IN DUCT BEHIND REGISTER OR GRILLE (ARROW INDICATES DIRECTION OF FLOW)	
TURNING VANES IN A SQUARE THROAT ELBOW	
TURNING VANES IN A ROUND THROAT ELBOW	
PLAN VIEW OF TRANSITION	
OFF-SET UP IN DIRECTION OF ARROW	

DUCT DIMENSIONS—FIRST FIGURE IS THE SIDE OF DUCT SHOWN (12 X 10)

MECHANICAL SYMBOLS (CONTINUED)

DESCRIPTION	SYMBOL
ACCOUSTICAL LINING INSIDE INSULATION	
BRANCH TAP IN DUCT	
SPLITTER FITTING WITH DAMPER	S.D.
VOLUME-DAMPER	V.D.
BACKDRAFT DAMPER	BDD
ACCESS DOOR IN DUCT 10"X 10" SIZE	10 x 10 A.D.
PNEUMATIC OPERATED DAMPER	P.O.D.
THREE-WAY VALVE	
PRESSURE REDUCING VALVE	
PRESSURE RELIEF VALVE OR SAFETY VALVE	
SOLENOID VALVE	
PIPE TURNED UP (ELBOW)	
PIPE TURNED DOWN (ELBOW)	
TEE (OUTLET UP)	
TEE (OUTLET DOWN)	

MECHANICAL SYMBOLS (CONTINUED)

DESCRIPTION	SYMBOL
BACKFLOW PREVENTER	B.F.P.
UNION	
REDUCER	
CHECK VALVE	FLOW
GATE VALVE	OR
GLOBE VALVE	OR
BALL VALVE	
BUTTERFLY VALVE	
DIAPHRAGM VALVE	
ANGLE GATE VALVE	
ANGLE GLOBE VALVE	
PLUG VALVE	

PIPING SYMBOLS

DESCRIPTION	SYMBOL
LOW PRESSURE STEAM	LPS
LOW PRESSURE CONDENSATE	LPC
PUMPED CONDENSATE	PC
FUEL OIL SUPPLY	FOS
FUEL OIL RETURN	FOR
HOT WATER SUPPLY	HWS
HOT WATER RETURN	HWR
COMPRESSED AIR	A
REFRIGERANT SUCTION	RS

PIPING SYMBOLS (CONTINUED)

DESCRIPTION	SYMBOL
REFRIGERANT LIQUID	RL
REFRIGERANT HOT GAS	RHG
CONDENSATE DRAIN	CD
FUEL GAS	G
CHILLED WATER SUPPLY	CWS
CHILLED WATER RETURN	CWR

PLUMBING SYMBOLS

DESCRIPTION	SYMBOL
METER	M
SPRINKLER PIPING	S
SPRINKLER HEAD	
FLOOR DRAIN	F.D.
CLEAN-OUT	C.O.
TUB	
TANK-TYPE WATER CLOSET	
WALL-MOUNTED LAVATORY	
URINAL	
SHOWER	
WATER HEATER	WH
MANHOLE	MH

PLUMBING SYMBOLS (CONTINUED)

DESCRIPTION	SYMBOL
WALL HYDRANT	
YARD HYDRANT	Y.H.
FLUSH VALVE WATER CLOSET	
COUNTER-TYPE LAVATORY	
KITCHEN SINK (DOUBLE BOWL)	

PLUMBING PIPE SYMBOLS

DESCRIPTION	SYMBOL
SOIL, WASTE OR DRAIN LINE	
PLUMBING VENT LINE	
COLD WATER (DOMESTIC)	
HOT WATER (DOMESTIC)	
HOT WATER RETURN (DOMESTIC)	
FIRE LINE	F
FUEL GAS LINE	G
ACID WASTE LINE	AW
VACUUM LINE	V
COMPRESSED AIR LINE	A
BACKFLOW PREVENTER	BFP
GATE VALVE	OR
GLOBE VALVE	OR
CHECK VALVE (ARROW INDICATES DIRECTION OF FLOW)	

PLUMBING PIPE SYMBOLS (CONTINUED)

DESCRIPTION	SYMBOL
UNION	
PIPE TURNED DOWN	
PIPE TURNED UP	
TEE OUTLET UP	
TEE OUTLET DOWN	
TEE OUTLET TO SIDE	
REDUCER	
PIPE SLEEVE	

ELECTRICAL SYMBOLS

DESCRIPTION	SYMBOL
SINGLE CONVENIENCE OUTLET	1
DOUBLE CONVENIENCE OUTLET	
CONVENIENCE OUTLET OTHER THAN DUPLEX, TRIPLEX, OR DOUBLE DUPLEX, ETC.	3, 4
SPECIAL PURPOSE OUTLET (DESCRIBE IN SPECIFICATIONS)	
FLOOR OUTLET	
FAN OUTLET	F
GENERATOR	G
SINGLE-POLE SWITCH	
DOUBLE-POLE SWITCH	2
THREE-WAY SWITCH	3
FOUR-WAY SWITCH	4

ELECTRICAL SYMBOLS (CONTINUED)

DESCRIPTION	SYMBOL
KEY-OPERATED SWITCH	K
SWITCH WITH PILOT LIGHT	P
WATERPROOF SWITCH	WP
FUSED SWITCH	F
GANG OUTLET	R
SWITCH AND CONVENIENCE OUTLET	S
JUNCTION BOX	J
LAMP HOLDER	L
WALL-MOUNTED LAMP HOLDER	L
CLOCK OUTLET	C
LIGHTING PANEL	
POWER PANEL	
BRANCH CIRCUIT	
HOME RUN TO PANEL (NUMBER INDICATES THE NUMBER OF THE CIRCUIT BREAKERS IN PANEL)	1
BRANCH CIRCUIT (HASH MARKS INDICATE NUMBER OF WIRES)	
MOTOR	OR M
ISOLATING SWITCH	
BELL	
OUTSIDE TELEPHONE OUTLET	
FIRE ALARM BELL	F
FIRE ALARM STATION	F
TRANSFORMER	T

ELECTRICAL SYMBOLS (CONTINUED)		ELECTRICAL SYMBOLS (CONTINUED)	
DESCRIPTION	**SYMBOL**	**DESCRIPTION**	**SYMBOL**
FUSED ISOLATING SWITCH		TELEPHONE SWITCHBOARD	
INSIDE TELEPHONE OUTLET		BATTERY	
		GROUND POINT	

INDEX